The CAD Manager's Handbook

"Draw from Experience" to build your AutoCAD® library

COMPREHENSIVE AUTOCAD

- Stellman/HARNESSING AUTOCAD
  (Release 12 with Simulator® Tutorial Disk Package) — ISBN 0-8273-5917-9
  (Release 12 with Project Exercise Disk) — ISBN 0-8273-5930-6
  (Release 11) — ISBN 0-8273-4685-9

- Fuller/USING AUTOCAD
  (Release 12) — ISBN 0-8273-5838-5
  (Release 11) — ISBN 0-8273-5344-8

- Kalameja/THE AUTOCAD TUTOR FOR ENGINEERING GRAPHICS
  (Release 12) — ISBN 0-8273-5914-4
  (Release 11) — ISBN 0-8273-5081-3

- Rubenstein/AUTOCAD: THE DRAWING TOOL
  (Release 11) — ISBN 0-8273-4885-1

- McGrew/EXPLORING THE POWER OF AUTOCAD
  (Release 10) — ISBN 0-8273-3694-2

AUTOLISP & CUSTOMIZING

- Kramer/AUTOLISP PROGRAMMING FOR PRODUCTIVITY — ISBN 0-8273-5832-6

- Tickoo/CUSTOMIZING AUTOCAD
  (Release 12) — ISBN 0-8273-5895-4
  (Release 11) — ISBN 0-8273-5041-4

ADVANCED AUTOCAD

- Grabowski/AUTOCAD FOR WINDOWS — ISBN 0-8273-5581-5

- Grabowski/THE SUCCESSFUL CAD MANAGER'S HANDBOOK — ISBN 0-8273-5233-6

AUTOCAD APPLICATIONS

- Miller/AUTOCAD FOR THE APPAREL INDUSTRY — ISBN 0-8273-5224-7

GENERIC CADD

- White/DISCOVERING GENERIC CADD 6.0/6.1 — ISBN 0-8273-5571-8

CAD REFERENCE

- Grabowski/THE ILLUSTRATED AUTOCAD QUICK REFERENCE
  (Release 12) — ISBN 0-8273-5839-3
  (Release 11) — ISBN 0-8273-4820-7

To purchase, contact your local bookseller or write directly to:
Delmar Publishers Inc, 3 Columbia Circle Drive, P.O. Box 15015, Albany, NY 12212

The Successful CAD Manager's Handbook

Ralph Grabowski

Delmar Publishers Inc.™

I(T)P™

NOTICE TO THE READER

Publisher does not warrant or guarantee any of the products described herein or perform any independent analysis in connection with any of the product information contained herein. Publisher does not assume, and expressly disclaims, any obligation to obtain and include information other than that provided to it by the manufacturer.

The reader is expressly warned to consider and adopt all safety precautions that might be indicated by the activities described herein and to avoid all potential hazards. By following the instructions contained therein, the reader willingly assumes all risks in connection with such instructions.

The publisher makes no representations or warranties of any kind, including but not limited to, the warranties of fitness or particular purpose or merchantability, nor are any such representations implied with respect to the material set forth herein, and the publisher takes no responsibility with respect to such material. The publisher shall not be liable for any special, consequential or exemplary damages resulting, in whole or in part, from the readers' use of, or reliance upon, this material.

Cover Credit: **Michael Speke**
Book Design: **Ralph Grabowski**

Delmar Staff
Publisher: **Michael McDermott**
Associate Editor: **Pamela Graul**
Project Development Editor: **Mary Beth Ray**
Production Coordinator: **Andrew Crouth**
Art and Design Coordinator: **Lisa Bower**

Allen County Public Library
900 Webster Street
PO Box 2270
Fort Wayne, IN 46801-2270

For information, contact:
Delmar Publishers, Inc.
3 Columbia Circle, Box 15015,
Albany, NY 12203-5015

Copyright © 1994 by Delmar Publishers, Inc.
The trademark ITP is used under license.
AutoCAD is a registered trademark of Autodesk, Inc. All other product names are acknowledged as trademarks of their respective owners.

All rights reserved. No part of this work covered by the copyright hereon may be reproduced or used in any form or by any means—graphic, electronic, or mechanical, including photocopying, recording, taping, or information storage and retrieval systems—without written permission of the publisher.

Printed in the United States of America
Published simultaneously in Canada by Nelson Canada,
a division of the Thompson Corporation.

1 2 3 4 5 6 7 8 9 10   xxx   99 98 97 96 95 94

Library of Congress Cataloging-in-Publication Data:
Grabowski, Ralph
 The Successful CAD Manager's Handbook/Ralph Grabowski
  p. cm.
  Includes index.
  ISBN 0-8273-5233-6
  1. Computer-aided design. 2. Electronic data processing departments—Management. I. Title
 TA174.G69 1993
 620'.0042—dc20   93-34297
                                    CIP

Brief Contents

Introduction xv

Section I
Introduction to CAD Management

1 **The Role of the CAD Manager** 1
   Reasons to avoid and adopt computer-aided design

2 **Selecting the Software and Hardware** 11
   Six cardinal rules for purchasing CAD hardware and software

3 **Overcoming Problems with CAD** 21
   Dealing with CAD after it is introduced to your office

Section II
Creating a CAD Standard

4 **Colors and Layer Names** 33
   Selecting a color and layering convention

5 **Symbols and Filenames** 47
   Creating a standard for filenames and subdirectories

6 **Fonts, Linetypes, and Hatch Patterns** 61
   Standardizing and scaling text fonts, linetypes, and patterns

7 **Setting Up Dimensions** 71
   A look at the complexity of dimension styles

8   **Preparing the Prototype Drawing** 81
     Setting up an office standard for eight CAD packages

9   **Writing a CAD Standard** 143
     The table of contents for the in-house CAD standard

Section III
Maximizing CAD Efficiency

10   **Input, Output** 153
     Scanning paper drawings and plotting drawings

11   **Working with a Service Bureau** 167
     Services and prices provided by service bureaus

12   **Data Exchange** 173
     The pitfalls of exchanging drawings with other CAD packages

13   **Networking CAD** 187
     Connecting CAD workstations together

14   **Maximizing CAD Efficiency** 195
     Seven areas where CAD packages can be customized

Appendices

A   **Gallery of CAD Software** 211
     Descriptions of 26 CAD packages

B   **CAD Software Sources** 239
     A listing of CAD software products from 152 vendors

C   **CAD Layer Standards** 255
     The layer standards based on AIA and CSI

D   **The AutoCAD Certification Exam** 263
     A sample examination for certifying CAD operators

Index 275

Table of Contents

Introduction xv
About the Author xxi

Section I
Introduction to CAD Management

1   The Role of the CAD Manager 1
    Why CAD? 1
       The Cost 1
       Fear of the Unknown 3
    Seven Reasons to Adopt CAD 3
       1. Faster Drawing Production 3
       2. Better Drawing Accuracy 3
       3. Design Analysis 4
       4. Cost Estimates 4
       5. New Services 4
       6. Client Presentations 5
       7. Market Competition 5
    Six Problems Created by CAD 5
       1. Money 5
       2. Education 6
       3. Changes to the Organization 7
       4. Support Staff 7
       5. Communications 8
       6. Limited Resources 8
    Planning the Transition to CAD 9
       Rule: Keep everyone fully informed
    Summary 10

2 Selecting the Software and Hardware 11
Selecting the CAD Software 11
   Rule: Use the CAD software your clients use
   The Dual-CAD Office 14
   Alone Again, Naturally 14
   Add-ons Determine the Primary Application 15
Selecting the CAD Hardware 16
   Rule: The software determines the hardware
   Rule: Avoid proprietary solutions; buy standard parts
   Rule: Purchase standards, not state-of-the-art
Upgrading Hardware 20
   Rule: Replace the CPU when the speed improvement is 3x
Summary 20
   Rule: Don't buy anything until its need is proven

3 Overcoming Problems with CAD 21
Controlling Management Expectations 21
The Return on Investment 23
   Return-on-Investment Calculation 23
   Send Managers to Trade Shows 24
   Competitive Pressure 24
When Management Doesn't Like to Spend on Training 25
   Post Tips on Network 25
   On-Line Help 26
   Forgetting the Training 26
   Training Management 26
   Monthly User Meetings 27
Give Engineers the Right Tools 27
Keeping Management in the Loop 28
The Cost of Revision 28
   Rule: Only upgrade when the need is proven
The Final Analysis 29
Summary 29

Section II
Creating a CAD Standard

4 Colors and Layer Names **33**
   Why Use Color in Drawings? 33
      The Case for Monochrome 33
      The Case for Color 35
      How CAD Works with Colors 36
   What are Layers? 38
      Layers in CAD 38
      How to Name a Layer 38
      Strategy #0: Do Nothing 39
      Strategy #1: The Simple Plan 39
      Strategy #2: The Plotter Plan 39
      Strategy #3: The Four-Step Plan 40
      Strategy #4: Do What Your Client Says 42
      Layer Name Standards 45
         Limitations of CAD Systems 45
         Conflicts of the Disciplines 46
         Human Quirks 46
   Summary 46

5 Symbols and Filenames **47**
   What are Symbols? 47
      Symbols in CAD 50
      Rule: Anything drawn twice should be turned in a symbol.
      How to Create a Symbol Library 49
         1. Draw the Symbol 49
         2. Decide the Orientation 50
         3. Select the Insertion Point 50
         4. Determine the Scale 50
         5. Appoint the Layers 51
         6. Add Attribute Information 52
         7. Name the Symbol 52
         8. Document the Symbol 53
         9. Store the Symbol 53
      Sources of Symbols 54
   External References 54
      Uses for External Drawings 54

Drawing File Names 55
    Project-Discipline-Drawing 55
    Discipline-Type-Detail-Sheet-Revision 56
    Description-Floor-Discipline-Sheet 56
    Externally Referenced Drawings 56
    CalTrans Filename Conventions 57
  File Extensions 59
Summary 60

6   Fonts, Linetypes, and Hatch Patterns 61

Text Fonts and Styles 61
  Text Speed 64
    Font Substitution 65
    Quick Text 65
    Fill Off 65
    Layers Off 65
Linetypes 66
  Hardwired vs Customized Linetypes 66
  One-Dimensional vs Two-Dimensional Linetypes 67
  Software vs Hardware Linetypes 67
  Scaling Linetypes 67
Hatch Patterns 68
Scaling Text, Linetypes, and Hatch Patterns 69
Summary 70

7   Setting Up Dimensions 71

Dimensioning with CAD 71
  The Anatomy of a Dimension 73
Dimension Parameters 74
  Dimension Line 75
  Extension Lines 75
  Arrowheads 76
  Dimension Text 77
  All Dimension Features 78
  Dimension Element Colors 78
  Scaling Dimensions 78
Using Dimension Styles 79
Summary 80

8 Preparing the Prototype Drawing 81
What is a Prototype Drawing? 81
    Preparing the Prototype Drawing 83
        1. Open a New Drawing 83
        2. Set Unit of Measurement 83
        3. Set Resolution of Measurement 83
        4. Set Style of Angular Measurement 83
        5. Drawing Origin and Extents 83
        6. Set Object Snaps and Drawing Modes 83
        7. Load Text Fonts and Define Text Styles 83
        8. Load Linetypes and Set Linetype Scale 83
        9. Create Layers 83
        10. Load Hatch Patterns and Pattern Scale 84
        11. Load Symbol Libraries and External References 84
        12. Set Dimension Variables and Scale 84
        13. Select Default Plotter and Set Plot Style 84
        14. Save Prototype Drawing 84
        15. Make Two Backups 84
    MicroStation 85
    AutoCAD 92
    CadKey 101
    Cadvance for Windows 107
    Drafix Windows CAD 114
    Generic CADD 121
    DesignCAD 2D 128
    AutoSketch for Windows 136
    Summary 142

9 Writing a CAD Standard 143
The Office Standards Manual 143
Title Page 143
Disclaimer Page 144
Table of Contents 146
Standard Office Library 148
After the Document is Complete 149
Summary 149

Section III
Maximizing CAD Efficiency

10 Input, Output **153**
  The Electronic Drawing 153
  Eliminating the Paper Trail 155
    Rule: The paperless office is as likely as a paperless bathroom
  The Input Options 156
    Archiving Drawings 156
    Converting Drawings 157
    Original Drawings Are Not Accurate 159
    The Two Editing Solutions 160
    The Partial Conversion Strategy 161
  The Output Options 162
  Summary 166

11 Working with a Service Bureau **167**
  What is a Service Bureau? 167
  Bureau Services 169
    Input 169
    Output 169
    Maintenance 169
  Service Bureau Pricing 170
    CAD Training Rates 170
    Digitizing Drawings Rates 170
    Plotting Drawings Rates 170
    Additional Charges 171
    Contract Considerations 171
  Summary 172

12 Data Exchange **173**
  Proprietary Drawing Files 173
  Drawing Exchange File Formats 175
    DXF 175
    IGES 176
    HPGL 177
    WMF 177
    DWG 178

      The Horror of Drawing Translation 179
          Unique Entities 179
          Entity Limits 181
          CAD Database Accuracy 181
          Text Entities 181
      DXF Via AutoCAD 183
          DXF Via Third-Party Software 185
      The Test Grid 185
      Summary 186

13   Networking CAD 187
      What is a Network? 187
      The Components of a Network 188
          Network Operating System 188
          Network Interface Card 189
          Network Cable 189
      Two Styles of Networking 189
          Central File Server 189
          Peer-to-Peer 190
          Multiple LANs 190
      The Complexity of Networks 191
          Traffic Copy Functions 191
          The LAN Manager 191
      Additional Benefits of a Network 192
      Features of a Network-Friendly CAD Package 193
      Summary 194

14   Maximizing CAD Efficiency 195
      Why Customize CAD? 195
      Parametric Symbol Libraries 197
      Screen and Tablet Menus 199
      Macros and Scripts 200
      Programming Languages 201
      Database Links 203
      Under Windows 204
      Optimizing CAD 206
      Summary 207

Appendices

A	**Gallery of CAD Software** **211**	
	Anvil-1000MD	212
	AutoCAD	213
	AutoCAD for Windows	214
	AutoSketch	215
	AutoSketch for Windows	216
	CadKey	217
	CadMax TrueSurf	218
	Cadvance for Windows	219
	DesignCAD 2D	220
	DesignCAD 3D	221
	Drafix Windows CAD	222
	EasyCAD	223
	FastCAD	224
	Generic CADD	225
	Generic 3D	226
	GFA-CAD for Windows	227
	Home Series	228
	Mannequin	229
	Mannequin Designer	230
	MicroStation	231
	PC-Draft CAD	232
	Planix	233
	3D Concepts for Windows	234
	TurboCAD Professional for Windows	235
	Upfront	236
	Vellum 3D for Windows	237
	VersaCAD Design	238
B	**CAD Software Sources** **239**	
C	**CAD Layer Standards** **255**	
D	**The AutoCAD Certification Exam** **263**	
	Drawing Exercises	266
	Index **275**	

Introduction

The client was on the phone, breathless: "We've installed our first CAD station. Now we want to begin our first drawing. Which colors do we use? What do we name the layers? How do we name drawing files and which subdirectories do we put them in?"

"Anything you want," I answered.

■

That's the problem with CAD (computer-aided design) systems—most are so flexible that you can create a drawing anyway you want; draw with any color, give layers and files any name, and store drawings in any subdirectory on the hard drive. The client was smart enough to know that his firm needed to define the layer, color, and file naming conventions *before starting the first drawing.*

My second, more realistic answer was to tell him the choices he had in selecting a system for organizing drawings. In the chapters of this book, you learn about conventions that help you get started on your first drawings. Other chapters introduce you to the problems and benefits of computerizing your firm's drafting boards.

The Audience for this Book

Some books written about CAD management are slanted toward a particular CAD package because the authors assume you use their favorite CAD package. In today's market, you cannot afford to be biased entirely toward Autodesk or Intergraph or any other vendor.

While a decade ago Intergraph and ComputerVision ruled the marketplace, the tables turned as Autodesk became the darling of the CAD set. Today, increasing compatibility with AutoCAD makes competitive CAD packages worth considering again.

For that reason, this book assumes you have a CAD system—*any* CAD system—or are looking into adopting a CAD system. The rules for running CAD in an office are much the same for any brand name—just the terminology is different. This book discusses CAD in a generic fashion, yet makes specific references to eight of the most popular CAD packages: AutoCAD, MicroStation, Cadkey, Cadvance, Generic CADD, DesignCAD 2D, Drafix CAD Windows, and AutoSketch.

The Organization of this Book

The Successful CAD Manager's Handbook is divided into three sections: an introduction to CAD management, developing a CAD standard, and maximizing the efficiency of your CAD system.

Section I: Introduction to CAD Management

The hardest part in making a decision whether to adopt CAD is the uncertainty created by the unfamiliar new technology. This chapter prepares your firm for the challenges and benefits of computer-aided drafting and design.

Chapter 1: The Role of the CAD Manager. While the CAD manager's role appears to be one of making sure your firm's CAD system is running smoothly, much of your time is spent justifying additional purchases of hardware and software. This chapter presents defenses against excuses, reasons, and justifications for and against implementing a CAD system.

Chapter 2: Selecting the Software and Hardware. With the hardware market rapidly changing, no book can dare recommend a hardware spec list and hope to remain current. Instead, this chapter provides guidelines for hardware purchases, with an emphasis on buying standard rather than proprietary solutions. For the purchase of software, this chapter emphasizes the importance of buying the CAD system (or systems) that your clients use.

Chapter 3: Overcoming Problems with CAD. Once CAD has been implemented in your firm, you find yourself facing a host of new problems that manual drafting never created. This chapter describes solutions to problems such as unrealistic expectations by management, the need for on-going training, and calculating CAD's return on investment.

Section II: Creating a CAD Standard

While a computer-based drafting system makes it easier to enforce drawing standards, it also becomes more important that standards be used when clients expect drawings in electronic form. This section describes how to create an office CAD standard.

Chapter 4: Colors and Layer Names. A number of semi-official organizations have created standards for naming layers and setting colors. This chapter describes three standards, as well as gives step-by-step instructions for creating your own in-house standard.

Chapter 5: Symbols and Filenames. Extensive use of symbols is the key to increasing CAD productivity in your office. Yet you must create a system of naming and organizing libraries of symbols. This chapter describes systems of naming symbols, drawing files, and using externally referenced drawings.

Chapter 6: Fonts, Linetypes, and Hatch Patterns. With the advent of CAD systems that can read PostScript and TrueType font files, drawings have access to thousands of fonts. This chapter warns against the "ransom note" look and shows how to create a standard with just one font and four sizes of text.

Chapter 7: Setting Up Dimensions. As with layers, international standards exist for the look of dimensions. This chapter describes how to create these standards in your CAD system.

Chapter 8: Preparing the Prototype Drawing. With all the elements in place, you can create a prototype drawing that ensures all your firm's drawings are made to the same standard. Specific step-by-step instructions are provided for eight CAD packages: AutoCAD, MicroStation, Cadkey, Cadvance, Generic CADD, DesignCAD 2D, Drafix CAD Windows, and AutoSketch.

Chapter 9: Writing a CAD Standard. Finally, you should also document your firm's CAD standard and place it in a three-ring binder. This chapter shows how to organize the standard in written form.

Section III: Maximizing CAD Efficiency

After setting the CAD standards for your office, you need to consider ways of making CAD work more efficiently.

Chapter 10: Input, Output. This chapter could have been named, "Working with Paper" since it describes how to get existing paper drawings into your CAD system, and how to plot CAD drawings onto paper. We also discuss whether you need to deal with paper at all by going "all electronic."

Chapter 11: Working with a Service Bureau. Consider your local CAD service bureau as an "office overload" center. This chapter describes the services provided by service bureaus, describes the charges for typical services, and gives advice on dealing with bureaus—including those located out of town.

Chapter 12: Data Exchange. If you're working with one CAD package but your clients demand drawings in another CAD package, this chapter tells you all about the trials and tribulations of data exchange: the types of data exchange formats, the problems you will face, how to overcome many of the problems, and when to not bother with drawing translation.

Chapter 13: Networking CAD. Probably the only thing tougher than implementing a CAD system in your office is adding the networking system. This chapter is a primer on networks: how to avoid installing a network, when to install one, which standards to select, and solutions to common problems.

Chapter 14: Maximizing CAD Efficiency. The fastest CAD system is not one with the fastest hardware; instead, the fastest CAD work is done by optimizing the CAD program itself and applying specialized software solutions. This chapter describes seven areas that can be customized in the eight CAD packages, along with six software products that speed up any computer.

Appendices
As additional information, the book includes the following appendices:

Appendix A: Gallery of CAD Software. A look at 33 CAD systems with a screen figure and a brief explanation of each.

Appendix B: CAD Software Sources. A listing of 270 CAD software products from 169 vendors.

Appendix C: CAD Layer Standards. A more detailed look at the CAD layer naming systems based on the AIA and CSI.

Appendix D: The AutoCAD Certification Exam. A copy of the examination given to test proficiency in AutoCAD.

Versions of CAD Software Covered

While this book is general to any CAD package, it targets computer-aided design software that runs under DOS and Windows on desktop computers. For specific examples, the following CAD packages are described:

Full-featured CAD

- AutoCAD Release 12 from Autodesk
- MicroStation Version 5 from Intergraph
- CadKey 6 from CadKey
- Cadvance 5 for Windows from IsiCAD

Low-cost CAD

- Drafix CAD Windows v2 from Foresight Resources
- Generic CADD 6.1 from Autodesk Retail Products
- DesignCAD 2D 6.0 from American Small Computer Business
- AutoSketch 1.1 for Windows from Autodesk

Conventions Used in this Book

Since this book is not specific to one CAD package, the problem of terminology arises. While all CAD packages agree on the use and meaning of the term "line," they all have different terms for just about everything else. Is it a "drawing" or a "design file"? Is it a "symbol" or a "block" or a "component" or a "group" or a "part" or a "cell"?

In general, I use AutoCAD's terminology throughout the book. Since it is currently the most pervasive CAD package, more readers will be familiar with AutoCAD terms than of any other CAD package. When speaking of a specific CAD program, then I use its terminology.

In the following pages, I use the following typographical convention:

- Command names are shown in **boldface**, such as the **BHatch** and **HS** commands

- Menu picks are separated by a vertical bar (|), as in **File | Open** and **Element | Line Style**

- Named keys are shown in square brackets, such as [Enter] and [Ctrl].

- Keys you press simultaneously are joined by a plus sign, as in [Ctrl]+C and [Alt]+[PrtSc]

Acknowledgements

The following people were extremely helpful to me in describing how CAD works in their office or by supplying a copy of their office's CAD standards manual: Donald Beaton, Brian Boatright, David Hiscocks, Michael Jackson, Michael Schley, and Andrew Smith.

A big thank-you to the vendors who supplied software for the book: American Small Business Computers for DesignCAD 2D v6; Autodesk for AutoCAD Release 12 and IGES translator; Autodesk Retail Products for AutoSketch 3.1 for DOS, AutoSketch for Windows, and Generic CADD 6.1; CadKey for CadKey v6; Computer Support for Arts&Letters; Foresight Resources for Drafix CAD Windows v2; IsiCAD for Cadvance 5 for Windows; Image-In for Image-In-Color Professional; Intergraph and Bentley Systems for MicroStation 4.0 and 5.0; and Microsoft Canada for MS-DOS 6.0 and Windows for Workgroups.

Thanks to Pamela Graul and Mike McDermott at Delmar Publishers for their patience in waiting on me to finish writing this book. Was it really two years?

Thank you to my wife Heather and my children Stefan, Heidi, and Katrina for making time for me to finish this book—between vacation trips, at that!

Ralph Grabowski
September 30, 1993
Abbotsford, BC

About the Author

Ralph Grabowski is a free-lance writer and independent consultant based in Abbotsford, British Columbia. He has written hundreds of articles about AutoCAD since 1985 and is the author of 13 books about computer-aided design.

Ralph holds the position of CAD Series Editor for Delmar Publishers, is a Contributing Editor to AutoCAD User magazine, and sits on the Review Board of InfoWorld magazine.

He is the former Senior Editor of CADalyst, the first magazine for AutoCAD users. Ralph received his B.A.Sc. degree in Civil Engineering from the University of British Columbia.

Contact the author by writing to PO Box 3053, Sumas WA 98295-3053.

SECTION I

Introduction to CAD Management

CHAPTER 1

The Role of the CAD Manager

As the manager in charge of CAD, your role is to ensure the maximum efficiency of the CAD system. While you will be plagued daily by nitpicky problems (such as dry plotter pens), your most important job is to provide justification for the implementation of CAD.

This chapter presents seven solid reasons to replace the drafting board with a computer. There are steps to prepare your firm before CAD appears in the office. Finally, this chapter alerts you to six problems created by the introduction of CAD.

Why CAD?

You will not find an accountant who isn't using a computer to calculate accounts, yet it is common to find engineering and architectural offices still dominated by manual drafting boards. The estimates are that only between 50% and 80% of such firms use CAD (computer-aided design and drafting). Those disciplines that are more easily computerized, such as PCB and electrical design, have a higher CAD penetration rate.

There are two primary reasons for avoiding the switch to computer-aided design:

The Cost
Replacing the drafting boards with a computer-based drafting station (called a "CAD workstation") costs roughly $5,500 per workstation. Once the initial system in place, you must add training costs, software revision costs, hardware upgrade costs, and system expansion costs. Chapter 2

Executive Summary

The CAD manager manages the computer-aided design system at your firm. But more importantly, the manager must handle the politics from upper management who may be suspicious of technology and drafters leery of the new "black box" plunked onto their desks.

Two Reasons to Avoid CAD
There are many reasons that can be invented to avoid implementing CAD. The two primary reasons are the (relatively) high cost and fear of the unknown.

Seven Reasons to Adopt CAD
By going digital, your firm gains advantages in the following areas:
- **Drawing production.** Complete drawings in shorter time.
- **Drawing accuracy.** Create drawings more accurately.
- **Design analysis.** Perform alternative studies, contextual fit, 3D modelling, schematic diagrams, and visualization.
- **Cost estimation.** Create quantity take-offs and database interfaces.
- **New services.** Computers create possibilities of additional sources of income for your firm.
- **Presentations.** Attract clients with impressive 3D animations.
- **Market competition.** Stay up-to-date—or keep ahead of your competition—with drafting and design technology.

Six Problems Created by CAD
After setting up computers on the drafters' desks, the CAD manager needs to get a handle on the new problems created by CAD:
- **Money.** After justifying the initial purchase of CAD, you must justify additional expenditures on CAD for upgrades and expansion on a continuing basis.
- **Education.** In addition to training the CAD operators, you must educate management and clients in the way of CAD.
- **Changes.** CAD will inevitably cause your firm to re-engineer the way it does business.
- **Support staff.** The number and type of support staff changes.
- **Communications.** Moving to a digital format creates compatibility problems with file formats from other software packages.
- **Limited resources.** As CAD takes hold and becomes successful, the money may not be there to add CAD resources, such as more workstations, networking, and supplemental analysis software. ∎

examines the cost breakdown for implementing a CAD system, while Chapter 3 presents a return-on-investment calculator.

Fear of the Unknown

To some, computers are a "black box" where things "happen" that the user doesn't understand. (Even experienced computer hackers blame some software bugs on interference from "photons from the Sun.") Even the keyboard can be a challenge equal to Mount Everest to those who didn't take typing in grade school. And since data is stored digitally on magnetic material, the user loses the all-too-familiar paper trail.

In the past, management's only use for computers was through a secretary who handled the electronic word processing. Now, with networking, electronic mail, and the pervasiveness of computerization, management can no longer avoid the computer. The challenges created by management being forced to deal directly with the computer creates stress in managers, which can lead to irrational decision making.

In addition to cost and fear, there are other reasons that makes firms avoid CAD. One common reason is that upper management was burned by implementing CAD a decade ago when workstations cost $100,000 each and ran on a processor that was simply not powerful enough for CAD.

Seven Reasons to Adopt CAD

Instead of concentrating on the reasons *not* to adopt CAD, let's look at seven reasons to do so. After careful analysis, you will find that the positive reasons outweigh the negative reasons.

Faster Drawing Production

After overcoming the learning curve, CAD allows drafters to draw faster. Access to libraries of commonly used parts eliminates repetitive drawing. Specialized drawing routines automate the process of drawing stairs, bolts, and parking lot lines. CAD is most efficient when existing drawings are reused.

Better Drawing Accuracy

More important than the speed gains, CAD is inherently more accurate than pencil drafting. Instead of the accuracy to the width of the pencil lead, CAD drawings are accurate to a dozen decimal places. Instead of eye-balling the middle of a beam, CAD finds the precise mid-point. Semi-automatic dimensioning measures the part, rather than the drafter supplying the distance.

As a side-effect to accuracy, CAD is inherently neat. Average drafters (such as myself) become excellent drafters. All lines, text, linetypes, and hatch patterns are drawn precisely the same. Tedious work (such as drawing text, linetypes, and hatch patterns) is automated by the computer.

Design Analysis

Just as computers made it easy to do "what-if" analysis via spreadsheets, CAD makes it fast for you to perform alternative design approaches to a project. Area, volume, and distance calculations are trivial when the data is digital.

Computer-generated visualization and animation let you see the product and look around it before the product is ever built. By creating a three-dimensional model early in the project lets you check building masses and space constrictions.

Engineers and architects can work together on projects when CAD workstations are networked together. By using reference files over a network, everyone on the design team sees each other's work.

Cost Estimates

By working with digital drawings, quantity take-offs change from tedious work to a semi-automatic process. Because it consists of mathematical vectors, a CAD drawing returns to you the length and number of items, the area and volume of objects, and allows near-instant updates.

When interfaced with a database or spreadsheet program, CAD performs cost and performance analyses. When interfaced with a word processor, CAD creates contact documents and reports.

New Services

After adopting CAD, your firm will find new sources of revenue. One architectural firm found its niche in providing facilities management services to clients. Since the architects had the construction drawings in electronic format, it was a natural follow-up to offer electronic management of the facilities after construction.

Another architectural firm adopted CAD, then discovered some of its employees had a natural flair for creating slick-looking computer-based presentations and animations. The result was a spin-off firm specializing in presentation work.

An engineering firm wrote a font-end program to control access to drawing files and plotting across a Novell network. After finding that other companies would rather spend the money than the time on similar software, the firm had modest success in marketing their CAD front-end.

Client Presentations

People like to be informed and entertained. When you implement CAD, you can use it to create presentations that help you win contracts. The CAD-based drawings become the source for publications, such as proposals, contracts, marketing pamphlets, and reports. Renderings and animations help the client more easily understand your work, solving potential problems and getting approvals earlier. The same can be used for getting financing and leading public hearings.

Market Competition

If no other logical reason works, then the threat of competition is effective in introducing CAD to an office. At one time, potential clients perceived a firm to be progressive if computers were used on the project. Today, your clients expect you to use computer on their projects. But don't just display a computer on a pedestal in the front office for all to see and ban computer use in the back office, as one architectural firm once did.

Increasingly today, perception no longer matters with clients. Large clients now routinely demand the delivery of drawings in electronic form on diskette or tape. That leaves your firm with no choice but to buy into CAD.

If your competitors are using CAD, then your firm needs CAD to remain competitive. Consider that productivity gains from CAD allows your competitors to bid less on proposals or bid the same and make a greater profit. One of the best things about computers is "value billing": your firm can charge a fixed amount at the current rate but perform the work in far less time—that makes profits soar.

Six Problems Created by CAD

After setting up computers on the drafters' desks, the CAD manager needs to work on the new problems created by CAD.

Money

CAD is capital intensive. This means that your firm pays a lot of money upfront on computer equipment and CAD software before seeing any payback from improved productivity. This represents a gamble for your firm: will the money invested now be returned later? Fortunately, the cost of a CAD station is one-tenth of the cost ten years ago, considerably reducing the risk.

Since computers have more moving parts than the drafting board, you have higher maintenance costs. Something breaks everyday. While intense competition has decreased the quality of computers, competition has also increased the length of warranties. Computers now have three-year warranties; graphics cards and modems have five- to seven-year warranties. For products no longer under warranty, the price of a company-wide on-site service contract can be negotiated to a reasonable level.

Software and hardware becomes obsolete with each software revision and hardware upgrade. Software upgrades are loaded with new features, which in turn make greater demands on the hardware. After your firm makes its initial purchase of CAD equipment, expect to pay the same price over again in the next six years for upgrading software and hardware (see Table below). The initial purchase price is roughly $5500 ($3500 for software and $2000 for hardware). The typical upgrade period is two years for software and three for hardware.

Upgrade Costs

Year	Software	Hardware	Cumulative Cost
1	$0	$0	$0
2	$500	$0	$500
3	$0	$2000	$2500
4	$500	$0	$3000
5	$0	$0	$3000
6	$500	$2000	$5500

After justifying the initial purchase of CAD, your toughest job will be justifying additional expenditures on CAD on a continuing basis.

Education

As CAD manager, you must arrange for the training of the CAD operators, which has a triple-cost associated with it: the cost of the training, the salary paid while in training, and the loss of productive work during the training period. Expect a training period of six weeks to three months, depending on the abilities of the drafter. The training period is a combination of formal teaching and on-the-job self-learning.

In addition to training the drafters, you must arrange for training the principles and clients of your firm in the ways of CAD. That doesn't mean they should know how to draw a line; rather, they should know what CAD can and cannot deliver. More on this in Chapter 3.

Changes to the Organization

CAD forces a firm to re-engineer itself. The lines of communication and the areas of responsibility change. I worked at an engineering firm in pre-CAD times where engineers drew the plans in pencil; the drafters merely traced over the plan with ink on mylar. In fact, the drafters were referred to as "tracers." Getting my drawings completed meant scheduling ink-tracing time with the head of our drafting department.

When CAD was introduced to the firm, the engineers didn't need the tracers and they didn't need to schedule time with the drafting department. Instead, the engineers produced the final drawings themselves. As a result, some drafters were trained as technicians and the remainder were let go.

Instead of scheduling time with the drafters, your CAD-based firm schedules time on the plotters. Fortunately, today's high-speed plotting technology has reduced plot times from 30-to-60 minutes (typical of a decade ago) down to 1-to-5 minutes. That means you no longer need to allow for *days* of plotting at the end of a project, but rather a single day.

With manual drafting, there is no question over the authenticity of the original drawing. With CAD, an original drawing doesn't physically exist: the drawing consists as magnetic patterns on a computer disk. Your firm needs to establish what constitutes the original CAD drawing. Here are approaches taken by three different offices:

- The original CAD drawing is kept on a floppy diskette and copied to the hard drive for editing. After the editing is finished, the drawing file is copied back on the diskette. The person with the diskette "owns" the original drawing.

- The original CAD drawing is kept "locked" in a specific subdirectory on the central file server. The drawing file can only be accessed via check-in/check-out software. The front-end program limits access to the drawing by one user at a time.

- A macro automatically places a date and time stamp on the drawing each time it is accessed. The most up-to-date drawing is found by checking the time stamp.

Support Staff

CAD only becomes successful is when at least one person is given (or takes on) the responsibility of CAD guru—a position different from CAD manager. Great efficiencies are realized by having in your office a person knowledgeable in computers and CAD software, preferably hired from within and not a recent B.Comp. graduate. This person must learn new techniques, teach users, attend conferences, read the journals, and log on daily onto CompuServe.

Communications

Going digital creates compatibility problems in file formats. While progress is being made to make the DWG file format a universal CAD standard, you still need to understand the DXF and IGES file formats used for exchanging drawings with other software programs.

In addition, you need to work with project consultants, clients, the blue printers, and the contractor solving *their* CAD-related problems.

Limited Resources

As CAD takes hold and becomes successful, the money is scarce for acquiring additional resources. There is a danger that upper management becomes complacent after the initial installation of CAD. They find the CAD system is paying for itself and decide to divert the budget to new priorities. The danger is that the CAD system and the operators are not able to expand their facilities and abilities. Some CAD managers find themselves in the unfortunate position of being able to justify new CAD resources only if employees were eliminated; the former salaries are used toward capital purchases.

The resources include the hardware and software upgrades mentioned earlier. But you will need to add networking (network interface card, wiring, hubs, a central file server, and network software), larger on-line storage devices (tape backups, optical, and CD-ROM drives), and additional output devices (thermal plotters, color plotters, and B-size laser printers) all of which demand more physical space and money.

Purchasing the CAD software is just the first step. You will find that you need to add third-party software—some of which is as expensive of the original CAD package—for specialized design work, such as structural and mechanical analysis. Even reasonably-priced utilities become very expensive when applied to 100 workstations.

Planning the Transition to CAD

There are many steps you take before the first CAD workstation lands on the drafter's desk. Perhaps the most important step in making the transition from pencil drafting to CAD is:

Keep everyone fully informed.

This motto will eliminate the agitation created by rumors and the uncertainty of employee's future employment possibilities. Employers who feel that the cloak of secrecy is the best wardrobe find later they are wearing the Emperor's new clothes as the office erupts in rebellion to the new technology.

Here is practical advice for planning the transition to CAD. These eight items are based on the experiences of several firms who made a successful transition from manual to computerized drafting:

- Realize that introducing CAD is truly a transition process. Your firm will not be able to afford a computer on every desk right away. At first, buy one CAD system and use it for one new, small project without a tight deadline. That process will give you more experience in CAD than reading all books ever printed on the subject.

- You can take advantage of the fact that only a few employees will initially have a computer. Find out which employees are excited about getting their hands on a computer and train them first. In most cases, their excitement transfers to fellow workers, lessening the resistance to computerization. Later, you will find that the keen employees become natural mentors to their peers.

- Announce that the profits gained from switching to CAD will be shared. CAD operators who earn the firm increased revenues are entitled a share in the profits, otherwise you risk loosing them to better-paying competitors.

- Assume that the entire office will eventually be 100% computerized—a computer on every desk. If some employees refuse to be involved with computers, give them two choices: (1) offer the employee training in computers at the company's expense, which all employees would receive anyhow, or (2) suggest that the employee begin the search for work elsewhere, offering company assistance in the search.

- Teach the employees that they are going to become "information managers" who are aided by the computer. Drafters, engineers, and architects will spend more time moving information around on the computer than drawing lines. Information in digital form increases your firm's efficiency because it becomes effortless to reuse the information.

- Design an in-house training facility. While off-site training may be necessary at first due to the lack of computers, it is more effective to train operators with your firm's own equipment. Give the training facility a catchy name, such as "N.D.A. University" if your firm's name happened to be Neenah, Delbert and Associates.

- Determine how files will be handled and how the drawings will flow throughout the office. Work out the interface between CAD and other software, such as word processing and networking. How will drawing notes written with the office word processor get into the CAD drawing? Remember to allow extra time for plotting drawings, time that isn't required under manual drafting.

- Begin work on office CAD standards, which forms the bulk of this book. Some firms have found that they gain more productivity from instituting consistent office standards on the CAD system than they did after converting to CAD.

Summary

The successful CAD manager eases CAD into the office by carefully planning the process and keeping everyone informed. While the primary problem you will face is finding money for upgrades to the system, above all remember that the computer is simply a more efficient tool for getting the firm's work done. The remainder of this book describes how to make your firm's use of CAD more effective.

CHAPTER 2

Selecting the Software and Hardware

The CAD system consists of two parts: software and hardware. The software gives CAD its personality; the hardware lets the software do its work. This chapter describes some of the factors you should consider before purchasing software and hardware and lists purchasing advice for every piece of hardware installed in a CAD workstation.

Selecting the CAD Software

With nearly 200 CAD packages to choose from, it becomes difficult to choose a package on the basis of feature lists. In any case, those feature lists can be misleading, as the following example illustrates.

One prominent CAD package, MicroStation, limits you to creating 63 numbered layers in a drawing file. AutoCAD advocates scoff at such a limitation because *their* CAD software has no limits on the number of layers, which can have names up to 31 characters long. MicroStation advocates counter that the 63-layer limit is easily overcome by employing MicroStation's superior reference file capability and, besides, AutoCAD lacks hierarchial layer organization, layer styles, and supplied layer standards.

The example shows that it can be misleading when you see the entries "63" and "Unlimited" in the Layers column of a product comparison table. While AutoCAD's layer structure seems more flexible (long names, no limits on number), MicroStation's structure is more powerful and user-friendly (hierarchy, styles).

> ## Executive Summary
>
> *There are six cardinal rules to selecting software and hardware for a CAD system.*
>
> **Before buying the software, consider this rule:**
> Rule #1: Buy the CAD package your clients use.
>
> There are three corollaries to this rule:
>
> - Add-on software determines the primary application package.
>
> - The most crucial aspect to working with clients is that you can exchange drawings electronically.
>
> - There is nothing wrong with being a dual-CAD office to satisfy the above rule and corollaries.
>
> **Before buying the hardware, consider these rules:**
> Rule #2: The software determines the hardware.
>
> Rule #3: Avoid proprietary solutions; buy standard parts.
>
> Rule #4: Purchase standards, not state-of-the-art.
>
> Rule #5: Replace the CPU when the speed improvement is 3x.
>
> Rule #6: Don't buy anything until its need is proven. ■

So which CAD package to choose? The answer, surprisingly enough, is simple:

"Use the CAD software your clients use."

Before I explain the reasoning behind that answer, lets look at some of the arguments CAD vendors use to entice you to their software.

- *Claim:* "*Our software has features that Brand X lack.*"

Truth: These days, any CAD package has most of the features of any other CAD package. The programming effort to make CAD work on desktop computers ended with the 1980s. In the '90s, the work of CAD programmers is to improve the user interface and to leap-frog the competition's feature set. If a feature is missing, you can either wait for the next update or program it yourself using the customization and extensibility features of the CAD software.

In my experience, many CAD vendors do not understand their competition's feature set. This occurs because vendors use different terminology and implement features in different ways. For example, AutoCAD has some "transparent" commands which are known as "immediate" commands in CadKey, while all commands are "non-modal" (immediate and transparent) in MicroStation. Thus, when one CAD vendor claims their competitor lacks a certain feature, it may be a mistruth stated unknowingly.

- *Claim:* "*Our software is supported by hundreds, even thousands, of third-party add-ons.*"

Truth: Most CAD packages have programming extensions that allow creation of add-ons for any disciplines, such as architecture, mechanical, electronics, and geographic information systems. The extensions are either created by yourself, or programmed by third-party programmers, or provided by the vendor. Although Autodesk boasts the largest number of add-on products of any CAD package, the truth is that many communicate via DXF, which means that they may work with almost any other CAD package as well. In addition, more third-party products work with multiple CAD products. At least one drawing viewer displays AutoCAD DWG, MicroStation DGN, and Generic CADD GCD files.

- *Claim:* "*We are compatible with AutoCAD.*"

Truth: No CAD package is compatible with AutoCAD, not even Autodesk's other CAD software, Generic CADD and AutoSketch. Just because the other CAD package reads and writes AutoCAD DWG files does not make it compatible. CAD vendors privately admit that they are currently aiming for compatibility at the plot level: that means a drawing plotted by AutoCAD and by another CAD package look the same. Getting back to the electronic format of the drawing, however, it will be many years before another CAD package is truly compatible with AutoCAD.

- *Claim:* "There is safety in numbers."

Truth: Market share is meaningless to the heads-down drafter. While AutoCAD currently has the largest market share in certain segments of the CAD market, it is useless to buy AutoCAD if your firm works for a Department of Transportation who requires drawings in MicroStation format. Do you remember products named CAD\Camera and VisiCalc? Competitors are spending large sums of money to unseat AutoCAD from its current #1 position, both in marketing and in product development.

Now that we have debunked common myths, let me explain the claim, "Use the CAD package your clients use." The most crucial aspect is that you be able to exchange drawings with your clients. After all, a primary reasons for moving from the drafting board to digital drawings is sharing.

While paper drawings are universal, electronic drawings come in a wide variety of file formats—probably as many as 150 different formats, one for every CAD package. Increasingly, clients demand that you supply them with the drawing electronically on diskette or tape. The client will be unimpressed if they cannot read your drawing files.

Translation is not the key. While it is possible to translate electronic drawings between different CAD formats, the translation is never 100%. Neither DXF, IGES, nor DWG solve the problem. Your client will be even less impressed to read a translated drawing and find missing and mis-translated entities. Later in this book, we spend an entire chapter on drawing translation.

The Dual-CAD Office

At one time, having two different CAD packages in an office was heresy. Today it has become common-place because of the rule, "Use the CAD system your clients use." If you have some clients using MicroStation and some that use Cadvance, install both in your office. You will probably balance the number of copies with the amount of billable work, such as 100 copies of MicroStation and 25 copies of Cadvance.

A second reason for the dual-CAD office is that you can take advantage of the competitive upgrade offers software vendors are making. In the summer of 1992, an AutoCAD user could get a copy of MicroStation, Cadvance, and Drafix CAD for about $1,100 in total—$4,600 in savings.

Alone Again, Naturally

If CAD is used strictly in-house at your firm, then my rule doesn't apply. Instead, buy the cheapest package that does most of what you want without needing any customization or add-on software. If your drawings are all 2D, then consider the low-cost (under $500) CAD software packages. For you, the 3D design and solids modelling features of the high-end packages is overkill.

Add-ons Determine the Primary Application

In some cases, the quality of the third-party application determines the CAD package you select. A consulting engineering firm originally used Anvil CAD due to an edict from head office. The local offices subversively used AutoCAD and the head office eventually change its mind. For the bulk of their design work, the firm continues to use AutoCAD. However, for roadway design the engineers were not happy with the add-ons available for AutoCAD. Then they discovered RoadWorks, which did everything they needed for road design but ran only with MicroStation. They spent some time analyzing the effect of having two CAD packages in their office. In less than a decade, the road design department switched from Anvil CAD to AutoCAD to MicroStation.

Selecting the CAD Hardware

Deciding on the hardware for your CAD software is easy when you keep these two rules in mind:

> *"The software determines the hardware."*

and

> *"Avoid proprietary solutions; buy standard parts."*

This book does not attempt to describe the pros and cons of operating systems and hardware platforms because my recommendations will become out of date too quickly. As I write this, the 150MHz DEC Alpha CPU is the darling chip, the Pentium is being independently benchmarked as not quite as fast as Intel claims, the NeXT computer has stopped production, DOS v6 is being eclipsed by expectations over DOS v7, Windows NT has just come out of beta, and the Unix forces are trying to get their act together.

At this point, however, I do recommend that your firm purchase the lowest-cost DOS compatible computer with the longest warranty (currently three years). Here are the parts you should consider:

- **CPU.** Get the fastest CPU on the market because within three years it will have become the slowest CPU on the market. CAD is very compute intensive and takes advantage of every MIPS (million instructions per second) a CPU delivers.

- **Speed doublers.** Some chips are designed to run twice as fast internally as externally. For example, the Intel 80486 DX2/66 chip runs 66MHz internally but is mounted on a 33MHz motherboard. IBM and other chip producers now have triple-speed CPUs that run at 99MHz. This allows you to purchase a computer based on a lower-cost motherboard with a higher-speed CPU. I consider double- and triple-speed CPUs a good buy.

- **Math Chip.** Until Intel released the 486DX CPU, you had to purchase the matching math chip, which almost all CAD packages demand. Now the math chip circuitry is typically included in the CPU, which represents a cost saving to end users. But be careful: Intel's 486SX-line of CPUs lack the math chip, as do certain 386- and 486-like chips from Cyrix and IBM.

- **Replacement CPUs.** Be careful with replacement chips. While all claim to speed up your computer, they may not work well with math-intensive software such as CAD. Non-Intel 486-based replacement chips sometimes lack the math chip circuitry required by nearly all CAD programs. Others have a smaller 1KB internal cache. Intel's line of Overdrive replacement chips offer "up-to-70%" speed improvement but at a relatively high price. Be aware that replacement chips do not work with all existing CPUs.

- **Memory.** Most computers sold today have 8MB RAM installed on the motherboard. That is adequate for running a CAD package under DOS. Your computer needs 16MB to run AutoCAD and MicroStation under Windows. Autodesk has stated that a minimum of 24MB will be required to run AutoCAD under Windows NT. For these reasons, ensure the motherboard can hold 32MB of memory. Some older motherboard designs have a bug that limits the maximum memory to 16MB RAM.

- **RAM and Disk Caches.** To let the CPU access RAM at top speed, make sure the motherboard has a 256KB RAM cache. A large cache costs more but doesn't speed up the computer; a smaller cache saves money but doesn't let the computer run as fast. However, any size of memory cache significantly speeds up the computer.

    To access the hard drive at top speed, get disk controller card with a disk cache, or else set up a disk cache in the computer's extended memory. Don't use a RAM disk: it uses up the RAM needed by CAD software and looses all its contents during a crash or power loss.

- **Graphics Board.** There is some confusion today over the interface for graphics boards: EISA (enhanced industry standard architecture), VBE (video bus extension), or PCI (personal computer interface). All attempt to improve graphics performance for the Windows operating system over the ISA (industry standard architecture) bus used in all IBM-compatible desktop computers. The problem is that the three interfaces are incompatible with each other; only EISA is upwardly compatible with ISA. Purchasing a computer with one of the three interfaces means you may not be able to upgrade at a later date due to the standard (EISA, VBE, or PCI) being superseded by a newer standard.

  For this reason, I recommend purchasing a Super VGA VESA-compatible ISA-bus board capable of 256 colors at 1024x768 resolution. As your firm's needs become more specialized, you can purchase additional graphics boards that often work with the existing board. For recording animations on video tape: an NTSC-compatible (or PAL, in Europe) board. To display photorealistic renderings: a 24-bit board that displays 16.7 million colors. For high-speed 3D rendering and rotation: a specialized 3D graphics board.

- **Hard Drive.** The price of hard drives has plummeted as users no longer purchase drives smaller than 100MB. Since CAD creates many large files, purchase the largest hard drive you can afford (at least 300MB), then add a disk doubler, such as Stacker, to increase the capacity to 600MB. Most drives are sold today with either the IDE (integrated drive electronics) interface or SCSI (small computer system interface) interface. The design of IDE makes it very easy to install a hard drive but be aware that 512MB is the largest hard drive capacity you can connect to the IDE interface card. To overcome the limit, connect a second 512MB drive or use the SCSI interface. The advantage to SCSI is that a single interface card handles up to seven devices.

- **CD-ROM Drive.** Although CD-ROM players are improving in speed and adopting a greater number of standards, the prices remain artificially high. I recommend buying the cheapest CD-ROM player since you will use it mostly for loading software and accessing data, neither of which require great speed or special standards.

- **Ports.** For CAD use, the computer should have two serial ports (for the digitizing device and plotter), two parallel ports (for the printer and some plotters), and possibly a SCSI port for the CD-ROM drive, tape back-up unit, scanner, and hard drive.

- **Modem.** I recommend an external 9600 baud modem for accessing CompuServe and other bulletin board services. External so that the modem can be used with more than one computer; 9600 baud to save on connect charges and long distance bills. The extra cost for the higher speed pays for itself in less than a year of use. Many modems work at 14,400 baud but few BBSs connect at that speed. For an extra $50, you can also have fax capability built in to the modem. Look for a unit compatible with the V.32bis modem and Class I/II fax standards.

- **Printer.** While the technology in dot-matrix and inkjet printers is improving, I recommend paying extra for a laser printer. It is whisper-quiet compared to the dot-matrix, is less hassle than the inkjet, and is much faster than either. Make sure the printer has enough memory to print a full page of graphics at 300dpi; most don't have sufficient memory and you have to add memory. The 600dpi printers need 6MB RAM to print a full page of graphics at top resolution.

  Consider the extra cost of a networkable high-speed B-size laser printer. Some consulting firms find these ideal for producing client-ready documents. Networkable means any computer can access the printer; high-speed means about 17 pages per minute. The extra cost is offset by eliminating print shop and courier charges.

- **Plotter.** While pen plotters continue to improve in features and drop in price, the market today belongs to the thermal plotters, which have taken over from electrostatic plotters. Thermal plotters use the same technology as the fax machine, but do it at double the resolution (400dpi), six times the size (up to 48" wide) and in color. Electrostatic plotters remain the fastest and most expensive solution to producing plots.

- **Cutters.** For physical plotting, output devices are available to cut. These range from simply attaching a knife to the pen plotter, to flat-bed laser beam plotters, to stereolithography. The knife cuts vinyl for making signs; the laser beam cuts cardboard pieces, which you glue together to form a scale model; the stereolithography creates actual 3D models out of liquid plastic or sintered metal.

- **Mouse.** The mouse has become rampant due to the popularity of Windows. Personally, I prefer the mouse since it is compatible with all software and takes up very little desk space. If the mouse has more than one button, all CAD packages let you define functions for the additional buttons.

- **Digitizing Tablet.** A digitizing tablet is traditional with CAD for two reasons: (1) it allows CAD's many commands to be laid out over a large area, and (2) it allows you to trace in drawings. With today's emphasis of on-screen menus, the first reason is loosing its grip. That leaves the second reason as the only reason for purchasing a tablet. Tablets are available in sizes ranging from 6"x6" to E-size. Don't bother with anything smaller than 12"x12".

- **Scanner.** More than any other piece of computer hardware, scanners come in the broadest range of sizes and capabilities. The market has products ranging from $99 monochrome hand scanners to $50,000 E-size color scanners. While it may seem natural to include a scanner in the initial CAD system purchase, delay the purchase until you've proven your office needs the scanner.

    In any case, scanning service bureaus abound. In an emergency, use your fax machine as a 200dpi scanner. Dial up a computer with a built-in fax card and send the drawing over the phone lines. Once received, the fax file can be converted to any common raster file format.

When purchasing any piece of computer hardware, follow this rule:

"Purchase standards, not state-of-the-art."

While the state-of-the art might tempt you with a speed gain, it becomes completely useless in less than two years. I recall horror stories of CAD users spending thousands of dollars on purchasing state-of-the-art graphics boards. Within a couple of years, device drivers are no longer provided by the vendor for the latest version of CAD software. The board becomes techno-junk before the firm can write it off from income taxes.

The primary advantage to DOS-standard computers is that parts are interchangeable. It really doesn't matter who you buy a computer from because chances are the vendor will go under within a couple of years. If the computer needs repair, no problem. Just cannibalize parts from another machine: the power supply, floppy drives, ISA interface cards, multi-sync monitor, and IDE hard drive are interchangeable among all machines.

Sometimes we are lucky to have the standard be the state-of-the-art. This typically only happens when a single product captures more than 50% of the market. Such is the case with Intel CPUs and Hewlett-Packard laser printers. The newest version of product from both companies pushes the performance envelope yet is backward compatible with earlier versions.

Upgrading Hardware

When should you upgrade the hardware? Sometimes the decision is made for you when new software demands additional RAM memory or more hard disk space.

How long should you wait before replacing the computer? In the case of upgrading the computer itself, the decision is much harder since CPU changes are incremental. My rule-of-thumb for system improvement is:

"Replace the CPU when the speed improvement is 3x."

Speed improvements much less than 200% faster (which is 3x expressed as a percentage) are not particularly noticeable. In any case, the increase in CPU processing speed is such that the 3x improvement occurs roughly every three years. Thus you should tend to avoid add-on products (whether hardware or software) or replacement chip that advertise themselves as 10% faster or even 70% faster.

Summary

The final rule I leave with you is:

"Don't buy anything until its need is proven."

Often, computer equipment is bought "just in case we need it." It sits on the shelf, remains unused, and becomes obsolete in under 12 months. Unlike spare car parts, there is virtually no market for used computer parts. As proof, my six-year-old son asked me to buy him a never-used XT motherboard at a recent computer swap meet. The price I paid? $1.50, including the turbo 8MHz Intel 8088-8 CPU.

In this chapter, you learned important guidelines to selecting software and hardware for the CAD system. In the next chapter, you learn how to tackle some of the problems brought on by CAD.

CHAPTER 3

Overcoming Problems with CAD

"CAD is like adding an outboard motor to the drafting board," consultant Joel Orr said at a recent seminar. It is no longer heresy to vocalize the thought that replacing the drafting board with a computer for producing 2D drawings does not justify the cost of equipment and training. CAD justifies itself on remodelling, changing existing drawings, performing analyses, and creating three-dimensional drawings. In this chapter, we look at the problems created by unrealistic expectations made of CAD.

Controlling Management Expectations

From management's viewpoint, the number two problem with CAD (we talk about the number one problem later) is that the learning curve isn't steep enough. Management may have been mislead by initial sales claims that assert instant productivity improvements by computerization of the drafting department. Possibly the worst demo is the fly-through animation (based on 3D CAD) with an upbeat music track and the salesperson's proclamation: "This is what *you* can do with our CAD system."

Reality is different. The rule of thumb is that it takes six months for a drafter to come up to speed on CAD. That means after six months of production work, the drafter is just as fast using the CAD station as the drafter was on the manual drafting board. After that, productivity gains depend on the amount of repetitive drafting involved in creating a drawing. Most disciplines find 2:1 productivity. That means drafters complete a drawing in half the time under CAD than manually.

Executive Summary

Some problems introduced by CAD are created by the vendors. Other problems are created by the uneven pace of technology. Here are some of the problems you may face, along with suggestions for overcoming them:

- Control management's expectations, which are often raised by dazzling demos of photorealistic animations bringing 3D CAD drawing to life.

- Accurately calculate the return on investing in CAD, which may take as long as three years.

- Don't expect instant productivity gains, since it takes drafters as long as six months to get up to speed on CAD.

- Be aware of the technological trends; send management to regional and national trade shows.

- When calculating the cost of training the CAD operators, factor in an amount for sending management for introductory training.

- Give users the correct tools: some may only need a drawing viewer; others will need CAD *plus* analysis or rendering add-ons.

- Recognize that upgrades cost more than just the price of the upgrade: add-in the cost of upgrading related hardware and software.

- Make user group meetings easy to attend by providing lunch.

- And always remember that CAD itself is "work in progress."

Drawings with many symbols, such as electrical drawings, may find 4:1 productivity; drawings are finished in one-quarter the time it took with manual drafting. As for the animation... well, the salesperson probably didn't mention the hours of preparation and processing time. A smooth animation needs to display 30 frames per second and each frame needs minutes (or hours) to process—not to mention the tens of thousands of dollars for specialized video recording equipment.

The Return on Investment

The number one concern to management is that CAD help show a profit. There are a number of strategies that the CAD manager can employ to convince management. Some provide direct profit, other strategies involve indirect profit.

Return-on-Investment Calculation

This is what management *really* wants to know: how soon is the payback? The expenses in CAD are the cost of hardware, software, training, and maintenance. The income comes from faster drawing production (maybe—more later) and increased business (more later, too).

To help with this hard-core calculation, Autodesk has written a ROI (return on investment) calculator. Based on a spreadsheet model, you enter data relevant to your firm and the ROI calculator estimates the length of time it takes your investment in CAD to be paid back:

$5,000	Cost of Hardware
$3,750	Cost of AutoCAD Software
$35	Cost Per Labor Hour (including overhead and benefits)
180	Labor Hours per Person per Month
	(Assumes 22.5 days per month, 8 hours per day)
6	Training Time Required (In months)
	(Enter value between 0 and 12)
40%	Productivity LOSS During Training Time
	(Enter value between 0 and 1, Example: 0.50 = 50% Loss)
50%	Productivity GAIN After Training Completed
	(Enter value between 0 and 2, Example: 0.75 = 75% Gain)
	All Other Annual Savings(Costs):
$1,000	Year 1: Includes maintenance contracts,
$0	Year 2: Service bureau charges, upgrades
$750	Year 3: Costs, additional training, etc.
$0	Year 4
$750	Year 5

	Year 1	Year 2	Year 3	Year 4	Year 5
Hardware Cost	($5,000)	0	0	0	0
Software Cost	($3,750)	0	0	0	0
Productivity Loss	($15,120)	0	0	0	0
Annual Productivity Gain	$12,600	$25,200	$25,200	$25,200	$25,200
All Other Savings(Costs)	$1,000	$0	$750	$0	$750
Net Annual Gain(Loss)	($10,270)	$25,200	$25,950	$25,200	$25,950
Cumulative Gain(Loss)	($10,270)	$14,930	$40,880	$66,080	$92,030
Cumulative ROI	57%	163%	271%	377%	486%
Average Annual ROI	57%	81%	90%	94%	97%
Net Present Value @ 10%	$64,312				
Internal Rate of Return	245%				
Payback Period	Year 2				

The Figure shows a sample output from the ROI calculator. Based on the assumptions listed at the top, the payback period is two years. Naturally, the accuracy of the result is only as good as the assumptions of costs.

Send Managers to Trade Shows

While CAD-specific magazines provide monthly news about CAD, they do not tell about everything that's happening in the industry. The best source of information are industry-specific trade shows. At the larger national and regional shows, firms display their latest software and hardware products.

In addition to products, most trade shows have conferences. These seminars provide two kinds of information: (1) tips on how others are using the current technology; and (2) trends and announcements on future technology coming down the pike.

By sending department managers to trade shows, your firm is fully aware of how the industry is developing and transforming as technology continues to have an impact. The cost of this knowledge is travel and accommodation at the trade show, the price of the conference, and the two or three days of lost time in the office.

Competitive Pressure

In the end, the only reason some firms get into CAD is because of competitive pressure. As CAD becomes more pervasive, the pressure increases on uncomputerized firms. (Would you employ an accountant that didn't use a computer?) The pressure comes from three directions: from your clients, from your competition, and from within your own firm.

Increasingly, clients demand the delivery of drawings in electronic form. That doesn't mean faxing the finished drawings to the client, it means delivering the drawings in CAD format on diskette or tape.

Some firms get into CAD simply because their competition has. In this computer age, a firm looks foolish if they shun the use of computers. Waiting for the price-performance ratio to stop falling is a foolish strategy.

The low price of personal computers allows employees to computerize the firm from the grass roots. When employees bring their own computers in the back door, firms are forced to realize of the benefits of CAD.

When Management Doesn't Like to Spend on Training

Firms lack money, time and organization for CAD training. Here are some training tips that don't cost money. They do, however, take a small amount of time and organization:

Post Tips on Network

If your office is networked, the best solution is to post tips on the central file server. Tips are collected from CAD operators themselves, from magazines, downloaded from CompuServe forums, and from the FaxBack services provided by some software companies.

At time of writing, CompuServe offered these forums for CAD software:

CAD Forums on CompuServe

CAD Software	Forum Name
General CAD topics	LEAP, CADDVEND Section 1
Ashlar Vellum	CADDVEND Section 8
AutoCAD, AutoSurf	ACAD
AutoSketch	ARETAIL, ASKETCH
CadKey	CADDVEND Section 7
Cadvance	CADDVEND Section 4
ClarisCAD	CLARIS
DataCAD	CADDVEND Section 7
ESRI	CADDVEND Section 9
FastCAD, EasyCAD	CADDVEND Section 3
Generic CADD, Generic 3D, Home Series	ARETAIL, GENERIC
IBM CAD, CAD/One	CADDVEND Section 10
MicroStation	CADDVEND Section 6

CAD vendors sometimes maintain their own bulletin board system, where you can download tips, news items, and product bug fixes. Tips needn't be limited to CAD: include tips on working with word processors, spreadsheets, databases, memory management, and the operating systems.

The problem with a computer-based tip service is having access to a universal "tip" reader. Under Windows, the Help engine works with files formatted to RTF (rich text format) word processing format. Under DOS, the List.Com program acts as a speedy file viewer for files formatted in plain ASCII.

On-Line Help

As software improves, the on-line help provided by the CAD software improves. Windows-based CAD software usually includes on-line help that takes advantage of the Windows Help engine's features, including context-sensitive help, hot-links to related topics, and instant definitions. DOS-based CAD packages are slowly adopting many of the same features for their own help systems.

While on-line help is usually not as extensive as the printed documentation, it is much faster to access. With CD-ROM technology becoming common, electronic help files are more complete than printed documentation.

Forgetting the Training

Some training sessions try to cram knowledge into a single course of several days length. The thinking is that the operators learn all there is about CAD, then go to work in front of the CAD station filled with their recently acquired knowledge. The drawback is that students tend to forget the knowledge in about the time it took to cram it in.

The better alternative is to schedule short, infrequent sessions. A suggested schedule is four hours a day, twice a week. That gives the students a chance to apply what they've learned. Initially, neophyte operators work inefficiently but with each four-hour course added to their knowledge, the operators become experienced by building upon their previously-acquired knowledge.

Training Management

An engineering firm asked me to explain the problems of using two different CAD packages in their office. The meeting consisted of the computer manager, CAD operators from several disciplines, and a couple of upper management types. I spent a quarter of the meeting answering basic AutoCAD questions from the upper managers: "What's a polyline?" and "What are blocks?"

This shows the importance of giving managers some training in CAD. If they don't understand the technology and terminology, they cannot make intelligent decisions. Convince managers to attend training—even if it takes a threat from upper management. It's preferable, though, to invite managers to attend a seminar to avoid embarrassing them.

Consider having a company-wide technology training seminar, once a month. Schedule the seminar during lunch so that your firm doesn't loose productive work time. Have the firm provide a catered lunch as an easy incentive for everyone to attend. Use the meeting to work out issues: show what is possible and what isn't.

Monthly User Meetings

One engineering department has a centrally-located blue book. As users come across problems, they scan through the blue book to see if anyone else has come across the problem. If not, they write the problem down in the blue book for others to solve.

Once a month, the department has a "CAD Users" meeting. Held at lunch time, the meeting gives users an opportunity to talk about problems and solutions. Part of the meeting is dedicated to working through unresolved problems listed in the blue book.

Having users help each other out with CAD problems is probably the most cost-effective training method your firm can implement.

Give Engineers the Right Tools

Not every person in your firm needs a full fledged CAD station on their desk. While the price of a CAD station is much less than the $100,000 it cost in 1980, today's $5,500 price tag still is more than some users need.

Some may only need drawing viewing software with redlining capability, along with cheaper hardware. Others may need the full CAD station plus specialized analysis software.

A cost savings is the "floating software license," which works only on networked CAD stations. It is rare that all operators use the CAD software at the same time. In fact, you can safely assume that CAD software is used by 50% of your operators at one time. The floating software license lets you buy one physical copy of the software but allows more than one person to use it. If your firm purchases ten licenses, then up to ten users can access the software at one time. The software keeps track of how many operators are currently using it.

Keeping Management in the Loop

The advantage to the manual drafting board is that managers keep an eye on the drafter's progress by simply going for a walk down the aisle past the drafting boards. On every board, the entire drawing is laid out for the manager to instantly see the drafter's progress.

With CAD, that is no longer possible. The drafter is usually zoomed into a small part of the drawing or may have layers turned off. In this way, CAD inadvertently places a barrier between the drafter and the manager.

By networking all computers in your firm and giving managers drawing viewers, the managers can independently keep track of the progress of each drafter and drawing. Most drawing viewers display up to 25 drawings on one computer screen, making it easy to scan an entire project worth of drawings at a time.

The Cost of Revision

After the hardware, software, and training are in place, the day comes when the hardware needs to be upgraded, the software updated, and the training renewed. The decision to upgrade should be a strategic decision and not a knee-jerk reaction. The rule of thumb is:

"Only upgrade when the need is proven."

Here are some considerations to deciding whether to upgrade. Some software vendors produce an update just to generate additional revenue. MS-DOS v6, for example, contained nothing that wasn't already available from third-party vendors.

Other times, an upgrade can be downright dangerous. The original IBM AT came with a hard drive that crashed and lost data. As someone once said, "Your job is to produce drawings, not live on the bleeding edge." I know of firms who have seen no reason to upgrade from AutoCAD Release 10 or from MS-DOS v3.3. However, you may find the software vendor won't support software older than the most recent one or two versions.

Due to competitive pressures, upgrades these days tend to be meaningful. If the decision is made to purchase the upgrade, some companies save money by upgrading only a few stations. Most software and hardware is backwards compatible, letting older products use data produced by the newer product.

Another money saving strategy is to purchase every second upgrade.

Competitive upgrades between software vendors means that upgrading from two versions ago costs no more than upgrading from the most-recent version. Skipping every other upgrade saves 50% of the cost.

However, be aware of the additional costs associated with upgrades. If you upgrade your primary CAD package, you may be forced to upgrade all third-party add-ons as well. If you upgrade the hardware or operating system, peripherals may need new device drivers. In any case, your operators (and management) will need additional training.

The Final Analysis

CAD is still "work in progress." When we speak of the widespread popularity of CAD, we speak of computerized *drafting*. Drawing lines and circles has been automated to the point of being a trivial process. Gigabytes of digital drawings are produced every day.

The current challenge for CAD vendors and customers is dealing with the digital data. The second stage of CAD is computerized *design*. Design analysis software is expensive and doesn't do all that's required of it. An obvious example is linking a CAD drawing of a new roadway with all federal, state, regional and municipal requirements: planning, environmental, safety, and legal.

Summary

This chapter helped you anticipate the new problems brought on when your firm computerizes its drafting department. You learned strategies for overcoming the problems. In the next section of this book, you learn how to create CAD standards via the prototype or template drawing.

SECTION II

Creating a CAD Standard

CHAPTER 4

Colors and Layer Names

In the chapters of this section, you learn about conventions that help you get started in creating a system for organizing drawings. All CAD packages give you the freedom to use lots of colors and create many layers in a drawing. That leaves new CAD users in a quandary: What color coding system? Which layer naming convention? Why use colors and layers at all?

This chapter suggests the colors and layer names that can be used in a CAD drawing. It describes approaches to implementing a color and layer naming scheme for your firm's CAD system. The chapter ends with a description of three industry layer standards.

Why Use Color in Drawings?

Color is rarely used in manual drafting and you may question the thought of using color in an electronic drawing. Feel free to do so. Since the output from a CAD system is, in most cases, a black-white plot, you may not want to use color.

The Case for Monochrome

One structural engineering firm did just that. When it came time to upgrade their CAD hardware, the firm's principals got the CAD operators involved. Since the operators are the ones who use the equipment, it made sense to let them have a say in the selection of the graphics display. The operators decided on a very-high resolution monochrome display system for two reasons: (1) the final output would always be monochrome, and (2) they would be more productive since the extra-high resolution meant

Executive Summary

You should consider five approaches to creating a color and layer naming convention for your firm's CAD drawings, beginning with the infamous "Do Nothing Strategy":

Strategy #0: Do Nothing
When you use CAD infrequently, don't bother setting up a color and layer naming system.

Strategy #1: The Simple Plan
When you only use CAD in-house, create a simple color and layer system that makes sense for the kind of drawings you produce. At the very least, every drawings should have a layer for text, hatches and dimension.

Strategy #2: The Plotter Plan
If simple drawings are closely tied to a multi-pen plotter, you may want to match layer names to pen numbers.

Strategy #3: The Four-Step Plan
To create an intermediate-level in-house layer naming system, follow these four steps:
- Create a list of elements found in your firm's drawings
- Create a layer name for each element that conforms to your CAD package.
- Create sub-categories for the elements.
- Create a prototype drawing that contains the layers.

Strategy #4: Adopt a Convention
A number of organizations, such as the AIA and CSI, have developed layer-naming conventions suitable for their discipline.

If you work with clients, you may have no choice. Increasingly, large clients, such as CalTrans, dictate to consulting firms the use of colors and the names of layers, and files. ∎

less time-consuming zooms and pans. The money saved on loosing color allowed the firm to spend that money on higher resolution and bigger screens. The firm selected a graphics board that displayed 1600x1200 resolution with eight shades of grey on a 20-inch monitor.

The Case for Color

CAD programs use color for two purposes: (1) operator cues, and (2) controlling the plotter.

When a drawing contains colored elements, the colors provide cues to the operator, making users more efficient. It's easier to pick out a blue water pipe from a sea of red structural members than when pipes and steel beams are both drawn in black.

While the structural engineering firm decided against a color display, they still need to use color in drawings—even if the colors are not displayed or plotted. That's because all CAD systems use color to control the plotter's pens.

When it comes time to plot the drawing, the CAD program asks you to map color numbers to the plotter's pen numbers. To help you decide how to use colors in a drawing, here are some ways you can map colors to pen numbers:

- Most commonly, CAD operators match color numbers with pens of different thickness. For example, the operator matches color numbers 1, 2, 3, and 4 with the plotter's 000 pen, color number 5 with the 00 pen and color number 6 with the 0 pen. Everything drawn with color 1, 2, 3 or 4 is plotted with thin lines.

- When color output is needed, the CAD operator matches color number with the plotter's pen number. For example, color number 1 is matched with the red pen. Everything drawn with color 1 is plotted in red.

- For monochrome plotters, such as thermal and electrostatic plotters, the CAD color numbers match to line widths and shades of grey, rather than pen numbers.

- Naturally, you can map all colors to pen #1 for draft plots or for output to a monochrome laser printer.

Most CAD packages display up to 256 colors; Cadvance displays just 15, while AutoSketch is limited to 128 colors. While MicroStation displays 256 colors, you can select those colors from a total range of 16.7 million.

Most multi-pen plotters handle six or eight pens. Raster plotters, such as electrostatic, inkjet, and thermal plotters, handle 128 or 256 colors, which makes them useful for plotting renderings.

How CAD Works with Colors

CAD software doesn't work with color names but with color *numbers*. The software matches the number to a color. (Sometimes, the CAD software lets you specify a color name as a pseudonym for the number.)

Unfortunately, there are different ways of matching the number with the color displayed on the screen. Two of the most common systems are known as EGA and ACI. The Table below shows the color names for the first 15 color numbers for both systems:

CAD Color Numbering Systems

Name	EGA	ACI
Red	4	1
Light Red	12	9
Magenta (pink)	5	6
Light Magenta	13	14
Yellow	6	2
Light Yellow	14	10
Green	2	3
Light Green	10	11
Blue	1	5
Light Blue	9	13
Cyan	3	4
Light Cyan	11	12
White	7	7
Light Gray	15	15
Gray	8	8
Black	0	0

The EGA system, named after IBM's enhanced graphics adapter, is used by most CAD systems. AutoCAD and AutoSketch (but not Generic CADD) use Autodesk's proprietary ACI system, short for "AutoCAD

> ## Tip Number 1
> ## Gray Tones from Blueprinting
>
> To have the base map or existing facilities show up gray on blueprints, plot them using a green pen on the mylar. Plot all other features with a black pen. When the mylar is blueprinted, the green ink lets some light through, creating a grayshade effect. ■

Color Index." Beyond the first 15 colors, both systems break down. The EGA graphics board was only designed to display 15 colors at most. Autodesk's own display drivers match different colors to ACI numbers than Autodesk's ADI (Autodesk device interface) specification, which is used by third-parties to write display drivers for AutoCAD.

The CAD system usually assigns colors by layer, although in many packages you override the layer color to specify the color of individual entities in the drawing. The reasoning behind setting color by layer (rather than by entity) is that you usually draft common elements on a single layer, which logically have the same color.

When you match color numbers to pen numbers at plot time, you specify the physical pen location. When you assign color number 1 (which might be red) to pen #1, the plotter uses whatever pen is in holder #1,

> ## Tip Number 2
> ## Not Always 256 Colors
>
> Whether you can display all 256 colors depends on two factors: (1) your computer's graphics board, and (2) the device driver. Most of today's graphics board display 256 colors but typically only when the resolution is 1024x768 or less. At higher resolutions, the number of displayable colors falls to 16. The number of colors (and resolution) is dependent on the board's architecture and on-board memory. The *device driver* is memory-resident software that lets the CAD package communicate with the graphics board. The device driver acts as a translator, converting the CAD commands into instructions the graphics board understands. The device driver can limit the capabilities of the graphics board. This happens when the CAD package has no driver specific to your graphics board and you substitute a generic 16-color VGA driver. ■

whether it is red or black, thin or wide. Some plotters let you match logical with physical pens. This is useful for long plots, when you can match logical pen #1 to physical pen #1, #2, #3 and #4. When the first pen runs out of ink, the plotter switches to pen #2... and so on.

What are Layers?

If you are familiar with overlay drafting, then it is easier for you to understand layers in CAD. Overlay drafting lets you combine several mylar drawings to create a master blueprint. For example, to produce blueprints for the plumbing contractor, you combine the mylars of the site plan, the structural plan, and the plumbing plan. The plumber is not interested in seeing the furniture placement or the landscaping plan.

In CAD, layers let you have all elements of a project in a single drawing: site, structural, plumbing, furniture, and landscaping. You place each on a separate layer and then turn layers on and off to create plotted drawings as required. Turn off the furniture and landscaping layers to plot the drawing for the plumbing crew.

Layers in CAD

Most CAD systems allow you to create up to 256 layers in a drawing. Some are more limiting: MicroStation is limited to 63 layers, while AutoSketch supports a mere ten layers. At the other extreme, AutoCAD allows you to create an unlimited number of layers in a drawing.

Most CAD packages refer to layers by number, typically ranging from 0 to 255; however, it is becoming more common to let the user assign names to layers. Layer names might be limited to eight characters (as in Generic CADD) or up to 31 characters, as in AutoCAD.

How to Name a Layer

Most CAD systems are not helpful when it comes time to decide on layer names. Like a doctor who delivers your baby but doesn't name it, CAD vendors are happy to deliver the software but leave layer naming up to the customer.

AutoCAD, for example, allows an unlimited number of layers, yet it defines a single layer named "0" in its default prototype drawing—most unhelpful for the neophyte CAD user! A few CAD systems are helpful enough to include an optional layer naming convention. TurboCAD's default drawing defines three layers—Drawing, Construction, and Dimension—to get a new user started yet maintain some separation between elements. MicroStation is more helpful to the user by including the AIA's CAD layer guidelines as an optional layering convention.

Most CAD systems, however, present you with a blank sheet. For this reason, we present five strategies for naming layers, numbered 0 through 4. After reading through each, choose one that best fits your firm's needs.

Strategy #0: Do Nothing

When you use CAD infrequently, don't bother setting up a color or layer naming system. Draw everything in black and draw them on layer 0. For simple drawings, it's actually a waste of time to toggle colors and switch layers.

Strategy #1: The Simple Plan

When you use CAD in-house for simple drawings, create a simple color and layer system that makes sense for the kind of drawings you produce. At the very least, every drawing should have separate layers for text, hatching, and dimensions since they take the longest to redraw. When you are not working with these layers, turn them off to reduce the drawing's redraw time.

Call the layers "Text," "Hatch," and "Dim" and color them white. If your CAD package doesn't allows named layers, use layer numbers 1, 2, and 3. It's that simple. Don't use layer 0 since some CAD packages use that layer for special purposes.

Strategy #2: The Plotter Plan

You may find that drawing and plotting is easier if the layers match your plotter's pens. Here is a layer system that matches layers with the most common plotter pen widths:

Layer Names Based on Pen Widths

Pencil Drafting	Entity Color	Layer Name	Pen Width
Soft: 7B to 2B	White	1	Bold: 0.70mm or 0.03"
Medium: B, HB, F, H	Yellow	2	Medium: 0.50mm or 0.02"
Hard: 4H to 9H	Green	3	Extra Fine: 0.25mm or 0.01"
Medium: 2H to 3H	Any other color	4	Fine: 0.35mm or 0.015"

Some CAD packages (such as Generic CADD, MicroStation, and Cadvance) show line weights on the screen and match them to pens.

Strategy #3: The Four-Step Plan

Depending on the type of drawing your firm produces, you may want to create layers in addition to the basic three. Here are the four steps to designing an in-house layer-naming system:

1. Gather together a number of drawings and scribble down logical elements you find in all of them. We've already mentioned text, hatching, and dimensions. Other elements include electrical, plumbing, HVAC, LAN, NC tool path, plasma cutter, conifers, and shrubs.

2. Create a layer name that describes each element, such as "Plumbing," "Electrical," and "HVAC." Now shrink each potential layer name to fit your CAD package's naming limitation. For example, if your CAD package limits names to eight characters, "Plumbing" fits the limit but "Electrical" must to be chopped back to "Electric". If your CAD package uses numbers, cross-reference the layer number and its meaning. Post the list at each CAD station.

3. Now think about sub-categories. If your firm does renovation work, you need more than just one layer for plumbing. Drawings must show three kinds of plumbing: existing to remain, existing to be removed,

---

Tip Number 3
Wildcarding Layer Names

When we named the three plumbing layers, we began each with the same letters: "Pb..." and "Plumbing..." for a good reason. All CAD packages let you turn layer off and on individually or in groups by using wildcards:

- * (asterisk) matches a group of characters
- ? (question) matches a single character

You probably recognize these characters as the same wildcards used by DOS and Unix. You quickly turn off (or turn on) all the plumbing layers by referring to them as "Pb*" or "Plumbing*." The numbered layers are handled slightly differently. Since the *second* digit defines a similar group, use "?1" to turn off or on all plumbing layers. ■

and new. Each should be on its own layer. In a CAD package without the short name limitation (such as AutoCAD and MicroStation) you can afford to be descriptive: "Plumbing-remain," "Plumbing-remove," and "Plumbing-new".

To fit an eight-character limit, this step involves a second round of layer name shrinking: "PbRemain," "PbRemove," and "PbNew".

If your CAD package uses numbers for layer names, consider logical groupings. For example, all existing elements are drawn on the 10-group of layers, construction is drawn on the 20-group, and all maintenance is shown on the 30-layers. Within each group of ten layers, apply the individual disciplines, such as x1 for plumbing. You end up with: layer 11 is existing plumbing, layer 21 is plumbing to be constructed, and layer 31 is plumbing maintenance.

4. Finally, create a prototype (or "seed") drawing with the layers and colors you've defined. Each time a CAD operator begins a new drawing, the layers and colors are predefined. See Chapter 8 for more information on creating a prototype drawing.

The following lists are helpful in developing a layer naming convention:

Layer Names for In-House Drawings

Architectural	GIS	Civil	Mechanical	HVAC
Dim	Elec	Paving	Gear	Gridline
Grid	Escape	Street	Frame	Gridtext
Misc	Contour	Sidewalk	Boss	HvacExch
Text	Legend	Trees	Crk	Hvac-Sup
Symbol	Text	ExParking	Hand	HvacSuph
Ceiling	Title	NewParking	Jaws	MpFixture
Floor	Vent	Build	Web	St-Metal
Roof	Workings	Boundary	...	Title
Stairs	Matchline	DemoConcrete	...	...
Wall	...	DemoStreet	...	...
Door	...	NewConcrete	...	...

Layer Names for In-House Drawings
(Continued)

Bridge	Landscape	Irrigation	Generic	Mapping
Center	Amenities	Channel	0	Shore
Grid	Easement	Emitters	1	Stream
HRebar	PropLine	IdBox	2	Railroad
Leader	Walk	Irrig	3	Ferry
Rebar	Fence	Laterals	Hatch	Road
Soil	Shade	Mainline	Hidden	Street
Text	Lawn	MowStrip	Phantom	Bridge
...	Trees	Pipe	Solid	Powerline
...	Label	PipeSize	Text	Telephone
...	Curb	Turfhead	Dim	Hydro
...	Title	Valves	TitleBlk	Airport

Strategy #4: Do What Your Client Says

To be compatible with your clients, your firm may be forced to adopt their layering standards.

Since 1988, a number of organizations have formed committees to create a layer naming standard. A portion of the standards are listed here but see Appendix C, "CAD Layer Standards," which lists the layer standards in greater detail.

American Institute of Architects. The AIA's task force on CAD layer guidelines created a layer naming system for building design, construction, and facility management. The system is not intended for other kinds of CAD drawings, such as mapping, highway design, printed circuit design, process plant engineering and aircraft design.

The naming system uses multiple fields separated by hyphens, called "delimiters." The fields divide the layer name into logical sections. For example:

```
E-LGHT-EMER
E-LGHT-SWCH
```

are the layer names for electrical emergency lighting and light switches. Notice how the delimiters separate the layer name in the major category (E = electrical), group (LGHT = lighting) and subgroups, EMER and SWCH.

The standard allows users to tag on an additional field for a firm's own use, as follows:

```
E-LGHT-EMER-XXXX
```

Unfortunately the standard loses its coherency by allowing the delimiters to be replaced by modifiers. For example, to designate a layer that shows the temporary emergency lighting:

```
ETLGHT-EMER-XXXX
```

The task force arrived at a triple naming system to accommodate AutoCAD's 31-character layer names (called the "long-form" name), CAD packages with eight-character layer names (called the "short-form" name), and those limited to numbered layers, called the "numbered" name.

- The long-form name has four or five fields, such as E-LGHT-EMER-XXXX or ETLGHT-EMER-XXXX.

- The short-form name consists of the first two fields, such as E-LGHT or EXLGHT.

- The numbered name is assigned to specific layer names, such as 130 for the emergency lighting layer.

Construction Specifications Institute. Sometimes called the "16-Division" system, the CSI layer naming system uses six-digit numbers to separate a CAD drawing into layers based on construction materials.

For example, the following layer number represents existing emergency fixtures to remain on floor 3:

```
16531-3
```

- The first two digits, called the "Basic Divisions," represent 17 CSI divisions, numbered from 00 to 16. Each division represents a group of construction specifications, such as site work (02), concrete (03), and finishes (09). Here, division 16 represents electrical.

- The second pair of digits, referred to as the "narrow scope numbering," represent the details of each division, such as roads (0252), gypsum board (0925), and furniture (1262). Emergency fixtures are 1653.

- The fifth digit of the layer name is used to define the status of the construction material, as follows:

    xxxx1: Existing to remain.
    xxxx2: Existing to be removed.
    xxxx3: New construction
    xxxx4: Text
    xxxx5: Dimensions
    xxxx6: Hatching
    xxxx7 through xxxx9: User definable layer

    Thus, existing emergency fixtures to remain are placed on layer 16531.

- The final pair of characters are optional and designate the floor number of a multi-story project. The dash (-) is a delimiter, while the 3 represents the third floor.

For simple drawings, the CSI system encourages you to use just the first two digits as layer names. By picking ten of the 17 basic divisions, you have a workable layer system for AutoSketch.

The advantage to the 16-Division layer system is that it matches the CSI construction spec book. The logical structure makes it easy to toggle layer visibility with wildcard characters. The disadvantage is that the numbers are meaningless to a CAD operator unfamiliar with the CSI spec.

CalTrans Drawing Data Levels. The California Department of Transportation *CADD Users Manual* is an example of the guideline clients must use to ensure uniform procedures for creating roadway drawings.

CalTrans and its consultants create drawings by merging master drawings in various combinations. Since the drawings are based on Intergraph's software (such as IGDS, MicroStation, and RoadWorks), the layer names are numbered between 1 and 63. Intergraph, which produces IGDS and MicroStation, uses the word "level" for layers.

There is no advantage or disadvantage to the CalTrans layer system; it is simply the system your firm must use if producing drawings for California Department of Transportation.

The following Table lists the meanings designated by CalTrans for the 63 layers:

CalTrans Layer Designation

Level Number	Meaning
1 - 8, 12	Basic topographic map data.
9, 10	Sheet formats.
11	Not used.
13 - 30	Basic construction details.
31 - 35	Right-of-way data.
36 - 59	Data specific to type of plan sheet.
60	Non-geographical drawing data.
61	Final plan revisions.
62	As-built changes.
63	Not used.

Layer Name Standards

When committees study the problem of developing a single standard for naming layers in CAD drawings, they find the task impossible. There are three insurmountable problems to creating one all-encompassing standard:

Limitations of CAD Systems. CAD software implements a variety of layer naming systems. At one extreme, an AutoCAD drawing is capable of holding an unlimited number of layers with a name of up to 31 characters long. At the other extreme, an AutoSketch drawing is limited to ten numbered layers, 0 through 9. In between, most CAD systems use 256 numbered layers (0 through 255); if names are allowed, they are usually limited to eight characters.

A layer naming standard that caters to both extremes would be limited by AutoSketch. Clearly, ten numbered layers won't serve as any discipline's standard.

Conflict of the Disciplines. A layer naming system that meets the need of one discipline may not meet the need of another discipline. A mechanical drawing's layer names differ from a mapping drawing's layer names, which differ from a dress design's layer names.

While a CAD system, such as AutoCAD, can hold every layer name ever devised, most CAD systems handle only 256 layers. Thus, a layer standard may need to be split into disciplines to overcome the 256-layer limitation.

Human Quirks. Debate in the CAD journals has shown that users have different preferences in approaches to layer standards. Some prefer alphanumeric-based layer names; others prefer a numerical layer names. Some prefer a complex system that lets you separate out nearly any unique aspect of the drawing; some prefer a simple system that simply matches half-a-dozen layers plotter pen colors. Some prefer layers subdivided by sub-contractor; others prefer layers subdivided by building material.

Summary

In this chapter, you learned how to set up your first drawings with an in-house standard for colors and layers. In the next chapter, you learn how to create standards for symbols and how to set up file names and subdirectories for your drawings.

CHAPTER 5

Symbols and Filenames

The last chapter showed you how to set up your first CAD drawings with an in-house standard for colors and layers. In this chapter, you find out how to create standards for symbols and how to set up file names and subdirectories for your drawings. In some CAD packages, symbols and files are closely related—thus we look at both in this chapter. For example, a symbol external to AutoCAD is simply a drawing file and is closely related to externally referenced files.

What are Symbols?

Drafting often involves drawing many similar symbols. Whether transistors, valves, or windows, the same symbols are drawn many times over. Manual drafting has two solutions: use a green plastic template to guide the pencil or stick on photocopies.

With CAD, there is one solution: insert symbols that have been previously drawn. A single command places simple and complex symbols without drawing a single line! Figure 1, on the next page, shows a kitchen plan drawn with AutoCAD. The drawing was created by drawing the walls and placing symbols. Symbols in the drawing include the stove, counters, sink, and windows. A symbol can include descriptive text, such as "Cutting Block" and database information, such as the manufacturer, size, and price.

Once a symbol is placed in a drawing, it is easier to edit with symbols than individual entities. A single pick selects the symbol for editing, rather than picking all individual entities. If you update a symbol, all copies are updated.

Executive Summary

With a layer naming system in place, you go on to create symbol standards and drawing filename standards. Follow these nine steps to create symbols for CAD drawings:

1. Draw the symbol.
2. Decide its orientation.
3. Select an insertion point.
4. Determine the scale, full size or unit size.
5. Appoint layer 0 or a specific layer.
6. Add attribute information.
7. Name the symbol based on the CAD package's naming limitations, usually eight characters.
8. Document the symbol in a three-ring binder or electronically.
9. Store the symbol in a read-only subdirectory; keep a backup copy off-site. ■

Symbols in CAD

Most CAD systems allow you to place an unlimited number of symbols in a drawing. You repeat each symbol as often as required. In fact, using symbols is more efficient (uses less disk space) in CAD drawings since multiple occurrences of a symbol are referenced. If you insert a bathtub symbol three times, the drawing file contains one copy of the bathtub and three references to the symbol definition. In CAD, the rule of thumb is:

"Anything drawn twice should be turned into a symbol."

CAD systems differ in naming symbols. Most allow an eight-character name. MicroStation is limited to six characters, AutoCAD lets you use 31 characters.

Symbols are known by a wide variety of names: "blocks" in AutoCAD, "cells" in MicroStation, "patterns" in CadKey, "parts" in AutoSketch, "components" in Generic CADD, "groups" in Design CAD, and "symbols" in Cadvance.

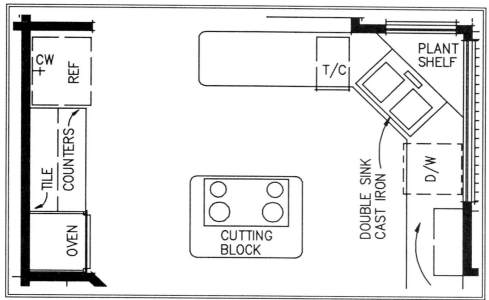

Figure 1: *This kitchen plan was quickly drawn by placing predrawn symbols.*

How to Create a Symbol Library

To make your firm's drafting more efficient, create a central library of symbols that all CAD operators have access to.

Just as you must create a standard for naming layers, you must create a standard for drawing and naming symbols. Unlike layers, however, no official organization has created a standard for creating symbols. This chapter presents methods you can use to create and name a set of standard symbols for your office. Here are nine steps to creating a symbol:

1. **Draw the Symbol.** When drawing the symbol, recognize that there are three types (see Figure 2):

Figure 2: *A CAD symbol is either literal (left), symbolic (center), or a reference (right).*

- **Literal symbols** that imitate physical objects, such as bathtubs and plants.
- **Symbolic symbols** that represent physical objects, such as a 45-degree piping elbow and a light switch.
- **Reference symbols** that indicate the location of non-physical objects, such as a North arrow and a contour.

2. **Decide the Orientation.** Symbols are usually placed in the drawing at different angles, but sometimes a symbol is placed at one particular angle more often than any other angle. For example, the North arrow symbol is more likely to point to the top of the drawing—thus draw it that way.

3. **Select the Insertion Point.** The *insertion point* (or "reference point") is the point in the drawing where the symbol is placed. The symbol is usually placed at the pick point (or "data point"). Like orientation, selecting an optimal insertion point now saves time later, although some CAD systems (such as Generic CADD) let you move the insertion point during symbol placement.

    Figure 2 shows the insertion point as a dot. The North arrow symbol has its insertion point at its tip to let you easily point it in the right direction. The bathtub symbol, on the other hand, has its insertion point where the tub attaches to piping. The piping symbol could have the insertion point at either end. For symbols that can have more than one logical insertion point, select one position as the standard for all similar symbols.

4. **Determine the Scale.** Literal symbols, such as bathtubs and plants, are drawn full size. The five-foot bathtub is drawn 60" long. The exception is the *parametric* symbol. This is a literal symbol that is scaled to size during placement. An example is a desk: save a single desk symbol as a 1" square. Upon placing the desk, you specify the x- and y-scale (and sometimes z-scale) factor. For example, to place standard-sized desks based on a unit block, use these scale factors (see Figure 3):

    - **24"x48" Desk:** x-scale = 24 and y-scale = 48
    - **24"x60" Desk:** x-scale = 24 and y-scale = 60
    - **30"x60" Desk:** x-scale = 30 and y-scale = 60

By using parametrics, a single desk symbol works for many different desk sizes. When working with large symbol libraries, parametrics save a lot of disk space.

Be aware that simple x- and y-scaling does not always work with parametric blocks. More complex symbols, such as bolts, have discrete design parameters, which can only be accurately scaled through the use of *look-up tables*. A look-up table lists discrete parameters, such as bolt diameter, head size, overall length, and thread length. A programming interface is usually required to access the lookup table and construct the symbol.

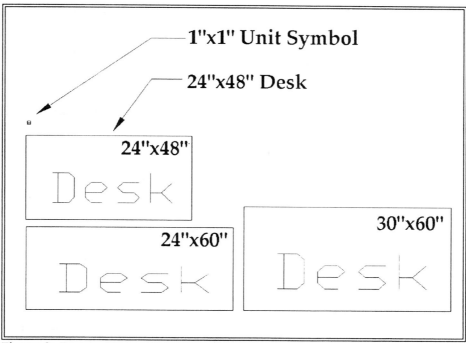

Figure 3: *Parametric symbols are scaled from a unit-size symbol.*

Symbolic symbols and reference symbols are drawn to unit size. Draw the pipe elbow and the North arrow to proportionally fit a 1" square. Later, when you place the symbol, you scale the symbol to the appropriate size.

5. **Appoint the Layers.** There are two kinds of layers on which to create the symbol:

- **Layer 0.** In some CAD packages, a symbol created on layer 0 has special meaning: when the symbol is placed in the drawing, it ends up on the current layer.

- **Any Other Layer.** A symbol created on any other layer is fixed to that layer: when the symbol is placed in the drawing, it ends up on the layer it was created and not the current layer.

For this reason, you must decide on your firm's layer system *before* you design the symbol library.

6. **Add Attribute Information.** In some CAD packages, attribute data can only be attached to symbols. *Attribute data* is information that is extracted from the drawing and exported to a database or spreadsheet program. For example, the bathtub symbol might include attribute data that describes the manufacturer, model number, and color.

    For some CAD packages, you define the attribute data while creating the symbol or you add the attribute data later. MicroStation 4 does not have attribute data; Version 5 allows you to attach tags to any entity in the drawing, including cells.

7. **Name the Symbol.** Finally, decide on the symbol's name. Most CAD packages limit the symbol name to eight characters. MicroStation limits cell names to six characters but lets you attach a long description.

    You can apply some of the layer naming procedures (discussed in the previous chapter) to naming symbols. In general, consider using a three-part filename consisting of the discipline name, part name, and an increment number, as follows:

    - **Discipline Name.** Reserve the first two characters of the symbol's name for the discipline. Here are some letter pairs to consider, based on the system used in Autodesk's old AEC package:

AR	Architectural
CD	Construction document
EE	Electrical
EQ	Equipment
FU	Furniture
LT	Lighting
MH	Mechanical, HVAC
MP	Mechanical, plumbing
TB	Title block
WS	Workstations

    - **Part Name.** The second pair of letters describe the part, such as WS**PS** for a PS/2 workstation and EE**LS** for a light switch.

    - **Increment Number.** Use the last two (or four) digits to define different models of the same part, such as WSPS*75* for a PS/2 Model 75 workstation and WWLS*0203* for a two-pole three-way light switch.

    Other symbol naming conventions you may want to consider include the manufacturer's part number or adapting the CSI 16-division system.

> **Tip Number 4**
> # When 31 = 8
>
> While AutoCAD allows 31-character names for blocks, *do not use more than eight characters!* There are two reasons to heed the warning:
>
> 1. The generous 31-character allowance only works with blocks stored in the drawing; as soon as you copy the blocks to disk (with the WBlock command), you are restricted to eight characters.
>
> 2. When using externally referenced drawings, AutoCAD uses up to 11 of the 31 characters for its own purposes. ■

8. **Document the Symbol.** After creating the symbol, document it. Most firms use a three-ring binder to document the symbol, others use electronic symbol librarians.

   Some firms document one symbol per page; others document four or nine symbols per page. Either way, here is the information you should include with each symbol:

   - Symbol name
   - File name (if different)
   - Library file name (if on a CAD system that stores symbols in a library file)
   - Description
   - Insertion (or origin) point
   - Orientation (if different from default)
   - Layer created on
   - Layer to be placed on (if different from creation layer)
   - Default plot scale (full size or unit size)
   - Attribute data (if any)
   - Drawn by
   - Last modified date
   - If a nested symbol, describe constituent symbols

9. **Store the Symbol.** When you are done, make two backups of the symbol library and make the collection of symbols available to everyone in your office. If your office is networked, place all symbols in a single read-only subdirectory, such as \CadSymbs.

Sources of Symbols

Frequently, symbols are available from other sources, saving your firm the work of creating its own. A number of manufacturers make their product catalog available on diskette at no charge. Usually, the products are saved in DXF format, often with attribute information included. Since symbols are very simple drawings, no data is lost translating from DXF to your CAD system's format.

Symbols are also available commercially from third-party programmers. (Some of the symbols illustrated in this chapter were created by CompugraphX.)

Clients may make their symbol library available to you, ensuring uniform drawing standards.

External References

An *externally referenced* drawing is a second drawing displayed by CAD software. You only view the referenced drawing and not edit it. You snap to entities and plot the referenced drawing. You can import (called "binding" by AutoCAD) all or part of the referenced drawing into the current drawing.

Externally referenced drawings are similar to symbols in that both eliminate duplication of effort. Do not confuse external references with paper space (found in ComputerVision, AutoCAD, and Cadvance), which allows you to view one drawing at two different scales.

The capability of CAD packages to support externally referenced drawings ranges from nil (for low-cost software) to complex. For example, AutoCAD and MicroStation can nest dozens of externally referenced drawings, where one drawing references another, which references yet another. MicroStation can self-reference a drawing (which AutoCAD can only do in paper space), allowing you to see parts of drawing in itself. While AutoCAD can clip rectangular portions of the referenced drawing, MicroStation can clip any shape. While MicroStation automatically updates externally referenced drawings, you must periodically use the AutoCAD's Xref Update command to ensure you are working with the most recent referenced drawings.

Uses for External Drawings

There are two primary benefits to using externally referenced drawings: (1) save disk space, and (2) workgroup design.

Instead of inserting a title block and border in every drawing, you reduce drawing file size by externally referencing the title block drawing. The information in the title block, of course, is entered in the current drawing using the referenced drawing as a template.

When your CAD workstations are networked, external referencing lets all members of the design team see each other's work. For example, the site plan is common to all construction drawings. The electrical, structural, and landscape designers externally reference the site plan. As the site plan is updated, the designers automatically receive the updates.

Drawing File Names

With the experience of naming layers and symbols behind you, naming the drawing files themselves should come much easier. As with symbol names, the DOS operating system limits you to an eight-character filename. To get around the eight character limit, make use of subdirectory names to separate files by projects.

Naturally, the filename of a drawing is closely related to the sheet number. If you decide to use a CAD drawing to produce a single plotted sheet, then the drawing's filename is the same as the sheet number. However, if a CAD drawing produces several different plotted sheets (via toggling the visibility of layers)—which is how CAD *ought* to be used—then the sheet number must differ from the drawing filename.

Here are examples of drawing filename conventions:

Project-Discipline-Drawing. An example filename is `60915A1.DWG`, where:

- 60591   is the project number
- A   is the discipline (A = architectural)
- 1   is the drawing number
- DWG   is the AutoCAD file extension

For the discipline section of the filename, you may want to employ the following discipline code letters:

A	Architectural plan
C	Civil and site plan
D	Demolition plan
E	Electrical plan
F	Food services plan
G	Graphics and signage plan
I	Interior design
L	Landscaping plan
M	Mechanical plan
P	Plumbing plan

S Structural plan
T Tenant plan
U User-defined plans

You may want to employ the following drawing type codes letters:

B Blocks (symbols) and external references
C Composite drawings
D Detail drawings
E Elevation drawings
N Enlarged plans
P Plans
S Section drawings
T Text schedules
W Wall sections

Discipline-Type-Detail-Sheet-Revision. An example filename is `AC06915B.DGN`, where:
- A   is the discipline
- C   is the drawing type (P = plan drawing)
- 06   is the detail number
- 915   is the sheet number
- B   is the revision number
- DGN   is the MicroStation file extension

Description-Floor-Discipline-Sheet. An example filename is `HQB2E001.VWF`, where:
- TS   is the drawing description (TS = tool shed)
- B2   is the floor number (B2 = second basement floor)
- E   is the discipline (E = electrical)
- 001   is the sheet number
- VWF   is the Cadvance file extension

Externally Referenced Drawings. Here are examples of drawing filenames for a large roadway design project using externally referenced files:
- RW01   right-of-way requirements, sheet 1
- PL03   plan, sheet 3
- GM05   road geometry, sheet 5
- PM07   pavement markings, sheet 7

CalTrans Filename Conventions. CalTrans uses three naming conventions for Intergraph design files:

Project Drawings. An example filename is D12345E01.DGN, where:
- D   the district code
- 12345   first five digits of the project expenditure authorization
- E   the sheet identification code (E = electrical)
- 01   the sheet number; letters A through Z represent sheet numbers 100 through 125.

Related Drawings. An example filename is DRHGKHSHR.DGN, where:
- D   the district code
- RHG   the operator's initials
- KHSHR   a drawing name assigned by the operator

Topographic Drawings. An example filename is D86725B27.DGN, where:
- D   the district code
- 86   the fiscal year of the HQ aerial survey contract
- 725   the contract order number
- B   the drawing code (B = base map)
- 27   the sheet number; sheet numbers 100 and higher eliminate the drawing code

CalTrans uses code letters, shown in the following Table, to identify the type of drawing in the filename:

CalTrans Drawing Codes

Drawing Type	Plan Sheet ID	CAD Filename
Sign plan	S	P
Retaining wall	R	Q
Sound wall	SW	R
Roadside rest	...	S
Planting and irrigation plan	HP	T
Lighting and signal plan	E	U
Revised standard plan sheets	...	V
Cross sections	Z	Z
Master design map	...	AA
Topographic base map	...	BB
3D terrain data	...	3D
Scanned drawing	...	CC
Digitized drawing	...	DD
Created drawing	...	EE
Project file directory	...	FF
Route adoption map	...	GG
Area of interest map	...	HH
Strip map	...	II
Freeway agreement map	...	JJ
New connection report exhibit	...	KK
PUC exhibit	...	LL
Geometric approval drawing	...	MM
Bridge site map	...	NN

File Extensions

CAD software uses a three-letter extension to the filename as a way for users to identify the source of the drawing files. Some CAD packages have more than one file format for storing drawings, such as AutoCAD. Other CAD packages have changed the extension for their design files, such as Generic CADD.

The Table below lists the filename extensions used by several popular CAD packages. Since the extension is usually a TLA (three-letter acronym), the meaning is also listed in the Table.

CAD Drawing File Extensions

CAD Package	File Extension	Meaning
...	IGS	*IGE*S (Initial Graphics Exchange Specification)
AutoCAD	DWG	*D*ra*W*in*G*
	DXF	*D*rawing Interchange *F*ormat
	DXB	*D*rawing e*X*change *B*inary
AutoSketch	SKD	*SK*etch *D*rawing
Generic CADD		
pre v5.0	DWG	*D*ra*W*in*G*
v5.0 +	GCD	*G*eneric *C*ADD *D*rawing
MicroStation	DGN	*D*esi*GN*
CADkey	PRT	*P*a*RT*
TurboCAD	TCP	*T*urbo*C*AD *P*rofessional
Cadvance	VWF	*V*ie*W F*ile
DesignCAD 2D		
pre v6.0	DC2	*D*esign*C*AD, v1
v6.0 +	DW2	*D*esignCAD *W*orkfile, v2

As with layers and symbols, you could consider using the CSI 16-Division system for naming drawings; or, you may be required to employ your client's file name system, such as the CalTrans system.

Summary

In this chapter, you learned how to create symbols to eliminate repetitive drafting. The chapter described how to create descriptive file names and subdirectories for drawings and symbols.

In the next chapter, you learn how to create standards for fonts, linetypes, and hatch patterns.

CHAPTER 6

Fonts, Linetypes, and Hatch Patterns

his chapter describes how text fonts, linetypes and hatch patterns are used in a CAD drawing. It explains the effect of scale factors and shows how to calculate scales.

Text Fonts and Styles

There is often confusion between text fonts and styles. A *text font* defines the basic font; the *style* defines variations on the font. Text font names include Txt, Simplex, and Sansserif. Text style names include oblique, large, and small (see Figure 1).

Traditionally, text in CAD drawings looked awful, often referred to as "stick fonts" due to the sparse look. The sparse look was a tradeoff for display speed. An example is the Txt font shown in Figure 1, above. Faster computer speed meant fonts could be smoothed out and even filled, as shown by the popular Simplex font and the triple-filled Roman Triplex font.

Today's CAD packages take advantage of the fonts available for desktop publishing. AutoCAD directly reads PostScript Type 1 PFB font files, while MicroStation reads PostScript and TrueType fonts. The Blueprint, Sansserif, and Romantic fonts shown in the Figure above are examples of PostScript fonts supplied with AutoCAD.

There are literally thousands of text fonts available for CAD and desktop publishing software. PostScript alone boasts over 10,000 fonts. In

Executive Summary

Text fonts, linetypes, and hatch patterns share a common problem: their size is a factor of the plotted output, rather than the drawing scale.

Text Fonts
Increasingly, today's CAD software is no longer limited to the vector-based "stick" fonts. Instead, many CAD packages read and translate any PostScript and TrueType font, many of which are available from free sources.

Still, the time text takes to redraw on the screen slows down the drafting speed. Here are four ways to speed up the redraw of text:

- Temporarily replace a complex font with a simpler font.
- Substitute text with "quick text," a rectangular box that outlines the text.
- Turn fill off to display just the font's outline.
- Turn of or freeze the layer containing the text.

Linetypes
- Some CAD packages hardwire linetypes, while others let the user define custom linetypes.
- Most CAD packages only allow one-dimensional linetypes (dots and lines); some allow 2D linetypes with text and symbols.
- Don't mix software linetypes (created by the CAD package) with hardware linetypes (drawn by the output device).

Hatch Patterns
- A few CAD packages hardwire hatch patterns, while most let the user define custom linetypes.
- Third-party vendors provides hundreds of hatch patterns at nominal cost. ■

recent years, the price of fonts have plummeted from several hundred dollars per font to free. For example, CorelDraw 4 includes at no extra cost 750 fonts in PostScript and TrueType formats.

The drawback to rich font selection is that it gives a "ransom note" look to drawings. Several dozen different fonts in one drawing looks

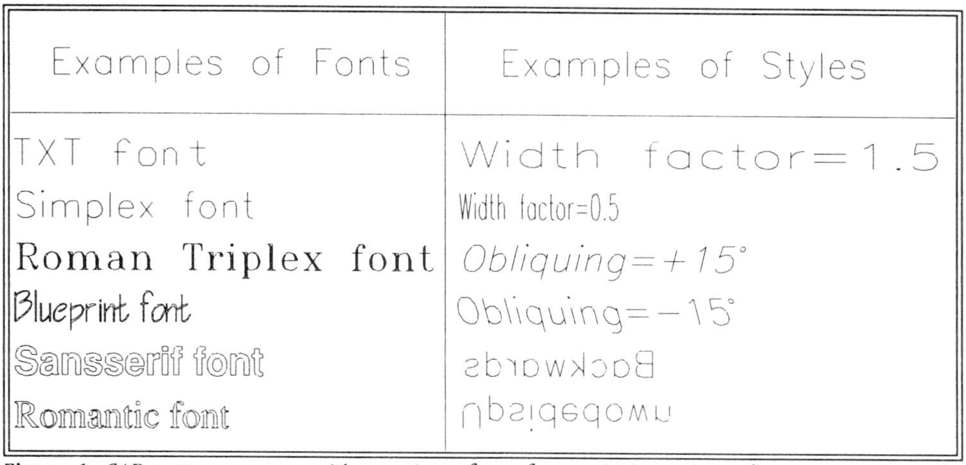

Examples of Fonts	Examples of Styles
TXT font	Width factor=1.5
Simplex font	Width factor=0.5
Roman Triplex font	Obliquing=+15°
Blueprint font	Obliquing=−15°
Sansserif font	Backwards
Romantic font	Upsidedown

Figure 1: *CAD programs come with a variety of text fonts, which can be defined in a variety of text styles.*

unprofessional and proves more difficult to read. Aim to work with a single font in a couple of styles and three sizes. Here is a suggested in-house standard (see Figure 2):

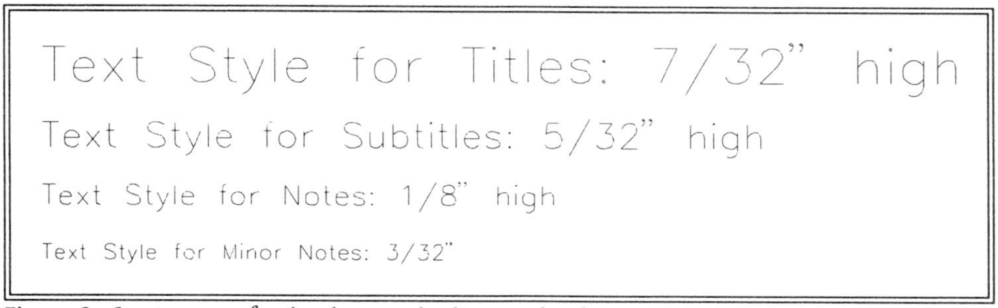

Text Style for Titles: 7/32" high
Text Style for Subtitles: 5/32" high
Text Style for Notes: 1/8" high
Text Style for Minor Notes: 3/32"

Figure 2: *Create a set of styles that standardize text height in drawings.*

- **Font:** Simplex
- **Style:** Based on size
- **Size:** 3/32" for small notes
  1/8" for notes
  5/32" for subtitles
  7/32" for titles

Some firms use 3/32" as the standard text size for notes. However, by going to the slightly larger (by 1/32") size, the text remains legible on drawings reduced to half-size.

Your clients may dictate other fonts and styles. For example, CalTrans expects the following text on drawings:

CalTrans Text Standard

Leroy Size	Height		Weight	Sample	Description
	Inches	Feet at 1:50 scale			
120	0.120"	6.00'	0	E	Restrictive text areas
140	0.140"	7.00'	1	E	Notes
175	0.175"	8.75'	2	E	Subtitles
200	0.200"	10.00'	3	E	Titles
240	0.240"	12.00'	3	E	Sheet title
290	0.290"	14.50'	4	E	Project description

The standard text font is Intergraph's #2 font, which is similar to the common Leroy font. For large titles, CalTrans suggests using Intergraph font #43, a filled sans serif block font.

Text Speed

The drawback to text is that it is vector intensive. The better looking the text font, the more vectors are needed to form the font. Fonts are filled with yet more vectors or a solid fill pattern, which take even more time to redraw. The Table compares the six CAD fonts illustrated in Figure 1

Tip Number 5
Place Notes on a Non-Plotting Layer

AutoCAD uses several layers for its own housekeeping purposes, including DefPoints. Since this layer does not plot, users sometimes place notes they don't want plotted on the layer. The notes are visible during editing but do not on the plot. ■

---

with AutoCAD's screen speed; the timings are measured in seconds for the equivalent of two pages of text.

Text Font vs. Screen Speed

Font Name	Regen	Redraw	QText
Txt	6.5	2.0	2.2
Simplex	10.5	3.6	3.1
Roman Triplex	28.5	10.1	7.6
Blueprint	38.6	14.5	9.3
Sansserif	44.1	17.0	10.4
Romantic	59.5	24.6	13.5

For reasons of speed, CAD software provides numerous methods to reduce the redraw time of text. The methods vary, depending on the CAD package, but here are some common strategies.

Font Substitution. Place text with its proper font. Then replace the text font with CAD package's fastest drawing font. Before plotting, swap the fonts again. Be careful that the two fonts are reasonable approximations of each other in terms of line lengths.

Quick Text. Some CAD packages have an option that replaces text with a rectangular outline. This lets you see the location of the text but takes very little redraw time. The drawback is that the rectangular outline usually is a poor approximation of the line lengths. (The timing shown in the Table, above, is the AutoCAD regeneration time for QText; the redraw time is too fast to time.)

Fill Off. CAD packages that can solid-fill text usually have an option to turn off the fill. Just before plotting, turn the fill back on.

Layer Off. Place text on their own layers, then turn off or freeze the text layers when not needed. This is the best option since it completely eliminates the text redraw time.

Text is used in many areas on a drawing. In addition to notes, text is used in dimensions, in the title block, and in visible attributes.

Linetypes

Linetypes were used in hand drafting to differentiate views: phantom and hidden lines show hidden parts. In some disciplines, such as mapping, linetypes denote data. For example, a line broken by dots indicates a property border while a line broken by the letter "T" designates a telephone line.

Traditionally, CAD packages implemented rudimentary linetype capabilities, then let the feature languish. It took eight years for Autodesk to fix linetype spacing in polylines; MicroStation didn't allow the user to customized linetypes until Version 5.

Hardwired vs Customized Linetypes

CAD packages either supply "hardwired" linetypes or allow customized linetypes. Hardwired are linetypes that the user cannot change. Generic CADD and AutoSketch, for example, have hardwired linetypes; up to Version 4, MicroStation had only hardwired linetypes. If your drawings need a linetype different from the few supplied by these CAD packages, you might be able to fake it or have to do without.

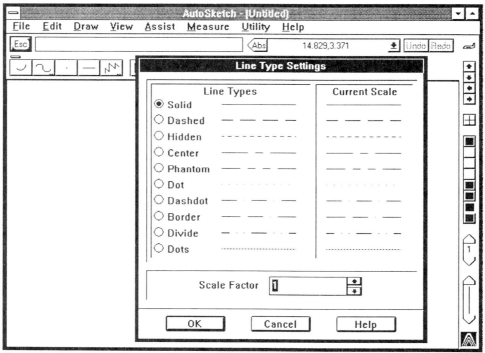

Figure 3: *Most CAD packages let the user define additional linetypes; some, like AutoSketch, have hardwired linetypes that the user cannot change.*

AutoCAD, MicroStation 5, and Cadvance, for example, allow customized linetypes. They include a small collection of linetypes and allow the user to add more linetypes.

One-Dimensional vs Two-Dimensional Linetypes

Most linetypes supplied by CAD packages are "one dimensional." That means the linetype consists of lines, dots, and space. A few CAD packages support "two dimensional" linetypes. This type includes text and shapes in the line.

MicroStation 5 provides the ability to create one- and two-dimensional linetypes. Third-party add-ons are available that let AutoCAD create 2D linetypes.

Software vs Hardware Linetypes

CAD software originally relied on the pen plotter to draw linetypes, since that was faster than having the CAD package generate linetypes. When the plotter draws the linetype, that is known as a "hardware" linetype; when the CAD package creates the linetype, that is known as a "software" linetype. Naturally, you don't use both the CAD software's and the plotter's linetypes on the same drawing.

To use hardware linetypes, you draw all lines in the CAD drawing with the continuous linetype. You differentiate lines by color. When it comes time to plot, you match CAD colors with the plotter's built-in linetypes.

Scaling Linetypes

When using linetypes, there are two scaling considerations: (1) the vector length, and (2) the plotted size.

When the CAD package draws the linetype, it begins at one end of the vector and draws the linetype pattern until it reaches the other end. Depending on the length of a vector (a line in the CAD drawing) and the scale of the linetype, the linetype might not show up or looks wrong.

If the vector is too short and the linetype is scaled too big, the linetype doesn't show up. Instead, the line appears solid because the vector doesn't have enough length for the CAD software to draw the lines, gaps and dots. The solution is to reduce the linetype scale or ignore the problem.

If the vector is long enough but the linetype scale is too small, the linetype also doesn't appear. However, in this case, the linetype makes its presence known: you see the line redrawn *v-e-r-y* slowly. This often happens when you are zoomed way out on a drawing.

Most CAD packages give you a choice in how to handle the linetype pattern at the two ends of the vector. The linetype starts at one end and stops in mid-pattern at the other end; the cleaner-looking alternative is to adjust the pattern at both ends to create a balanced linetype. The second choice is usually the default; in some drawings, you may want to turn the feature off. The plotted size of the linetype is discussed later.

Hatch Patterns

Hatch patterns are much like linetypes. A collection of patterns are supplied with every CAD package. Hatch patterns are never hard coded into the software, as are linetypes. The software documentation usually describes how to create your own custom hatch patterns, or you can purchase commercial collections at a typical price of about $1 per pattern—much cheaper than the time it would take to write your own.

Hatch patterns are like code that visually describes the properties of the objects in the drawing. Every discipline has its own set of hatch patterns with unique meaning. There tends not to be an office standard for hatch patterns; rather, each CAD operator refers to the patterning prescribed by the discipline, whether architecture, mechanical design, or mapping.

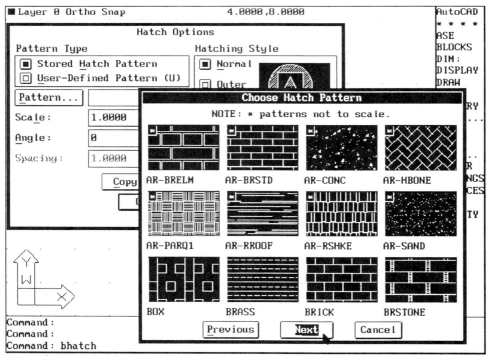

Figure 4: *While CAD packages include a selection of hatch patterns, you can design your own or purchase additional ones.*

Scaling Text, Linetypes, and Hatch Patterns

A difficulty with CAD is that you always work with two different scale factors: (1) most of the drawing is drawn full size at 1:1 scale, and (2) parts of the drawing are drawn at a different scale factor with the final plot size in mind.

Elements such as text, linetypes, and hatch patterns cannot be drawn full size. Instead, they must be drawn at a size that makes them legible when plotted. That means the scale factor must be determined before the first text, linetype, and hatch pattern is drawn.

Calculating the scale factor is not difficult. Setting it in the CAD drawing is perhaps a bit more difficult due to the confusing method the CAD software implements scale factors. For example, AutoCAD does not have one global scale factor; instead, you set scale factors independently for text, dimensions, linetypes, and hatch patterns.

The scale factor is usually determined by making normal text appear 1/8" tall. The scale factor is based on the drawing's plot scale. The scale factor can be set automatically via a setup routine (such as AutoCAD's MvSetup.Lsp program) or manually from the Table, shown on the next page. The scale factor table is based on the following formula:

$$\text{Text Height} = \frac{1/8\text{" Size}}{\text{Scale Factor}}$$

For example, the text height for a drawing plotted at 1:50 scale is calculated, as follows:

$$\text{Text Height} = \frac{1/8\text{"}}{1:50}$$

which equals

$$6.25\text{"} = \frac{0.125\text{"}}{0.020}$$

With the text scale determined for the drawing, the same factor can be applied to dimension text, leader text, hatch patterns, and linetypes.

Size of 1/8" Text

Plot Scale	Text Height
1/8" = 1'-0"	12"
3/16" = 1'-0"	8"
1/4" = 1'-0"	6"
1/2" = 1'-0"	3"
1" = 1'-0"	1-1/2"
1-1/2" = 1'-0"	1"
1' = 1000'	10'5"
1' = 500'	5'2.5"
1' = 250'	2'7.25"
1' = 100'	1'0.5"
1' = 50'	6.25"
1' = 10'	1.25
1' = 2'	0.25"
1' = 1'	0.125"
2' = 1'	0.0625"

The rightmost column in this Table shows the text height you supply as a parameter to the Text command in order to plot out text at 1/8" tall, based on the scale shown in the leftmost column. Using the formulae of the previous page, you can generate a similar text height table for the scale factors used by your office.

Summary

In this chapter, you learned how to create standards for text, linetypes, and hatch patterns. The chapter also described how to calculate scale factors for the plotted output. In the next chapter, you learn about setting up dimension standards.

CHAPTER 7

Setting Up Dimensions

Possibly the most complex aspect of CAD, dimensions involve seemingly endless options. Three of the CAD packages discussed in this book—MicroStation, AutoCAD, and CadKey—allow you to use multiple predefined dimension styles in their drawings. That makes it easy to have several standards on hand, such as ANSI, ISO, and DIN, particularly if your firm does international work. This chapter describes how dimensioning works in CAD and shows AutoCAD's approach to defining a dimension style.

Dimensioning with CAD

Dimensions are probably the most complex aspect of CAD. Although the typical dimension has only four components (extension line, dimension line, arrowhead, and dimension value), each can be expressed in a wide variety of means. Disciplines and regions each have their own way of drawing dimensions.

For this reason, almost all CAD packages provide a way to customize the look of dimensions, called the "dimension style." In this chapter, we look at how one CAD package, AutoCAD, sets dimension styles. We picked AutoCAD since its dimension parameters are complete yet explicitly presented via clearly laid-out dialogue boxes.

CAD packages offer different kinds of dimensioning and, naturally enough, give them different names. That can make it confusing to figure out the dimension capabilities of different CAD packages.

Executive Summary

All CAD systems include a facility for dimensioning parts in the drawing. The intelligence of the dimension facility varies greatly among CAD packages:

- Manual dimensioning forces you to draw all parts of the dimension yourself.
- Semi-automatic dimensioning lets you tell the CAD package the location of the two extension lines and the dimension line; based on those three points, the CAD software draws the dimension, including the measured distance.
- Automatic dimensioning lets you select an object and the CAD package dimensions it.
- Non-associative dimensioning means that after the dimension is drawn, it is independent of the object it measures.
- Semi-associative dimensioning means that when the *dimension* is stretched or condensed, it automatically updates itself.
- True associative dimensioning means when the *object* is stretched or condensed, the dimension is automatically updated.

Establishing a dimension style involves the tedium of setting dozens of parameters, yet only MicroStation includes predefined styles for ANSI (American), ISO (international), DIN (German), JIS (Japanese), Australian, an architectural style, and Intergraph's own INGR standard. ■

to help you differentiate between them, here are the six ways that a CAD system can dimension objects in the drawing. Dimensions can be created by the drafter, the CAD software, or a combination of the two, as follows:

- **Manual dimensioning:** you draw all parts of the dimension yourself.

- **Semi-automatic dimensioning:** you tell the CAD package where the two extension lines go and the location of the dimension line; the CAD software draws the dimension.

- **Automatic:** you select an object and the CAD package dimensions it.

Some CAD packages go one further on automatic dimensioning. A CAD package from Moscow dimensions every feature in the drawing with a single command! This is an interesting approach, since it is easier to erase a dimension than it is to place it.

Dimensions can be independent of the measured object, dependent on the object, or somewhere inbetween. This is called *associativity*, as follows:

- **Non-associative dimensioning:** after the dimension is drawn, it is independent of the object it measures.

- **Semi-associative dimensioning:** when the *dimension* is stretched or condensed, it automatically updates its value

- **True associative dimensioning:** when the *object* is stretched or condensed, the dimension is automatically updated.

If given the choice in a CAD package, always select automatic, true associative dimensions. For the most part, AutoCAD has semi-automatic, semi-associative dimensioning; only radial dimensions are true associative.

The Anatomy of a Dimension

To better understand why dimension styles are needed, let's look at their complexity. There are five basic types of dimension:

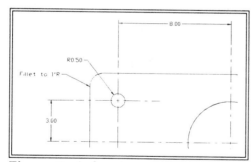

- horizontal dimension
- vertical dimension
- aligned dimension
- radial dimension
- leader

Figure 1: *A drawings contains five kinds of dimensions: horizontal, vertical, aligned, radial, and leader.*

Each type of dimension consists of four elements:

- dimension line
- extension line
- arrowhead
- dimension text

But each element has many variables. For example, the arrowhead might be an arrow, a dot, a slash, or nothing at all. The arrowhead's size, color, and layer can differ from the other dimension elements. Let's look at each parameter in detail as AutoCAD presents a series of dialogue boxes.

Dimension Parameters

AutoCAD lets you set dimension parameters in two ways, via:

- Dialogue boxes, via the DDim command.
- System variables, via the SetVar command.

We present both in the following sections.

Use the **DDim** command (or select **Settings | Dimension Style** from the menu) to display the opening dialogue box. The dialogue box displays a list of saved dimension styles on the left, and buttons for setting dimension parameters on the right (see Figure 2). While AutoCAD comes preconfigured with a reasonable set of dimension parameters (named "*Unnamed"), the package does not include any standards, such as ANSI, JIS, ISO, or DIN.

For the remainder of this chapter, we work through the contents of each button of the Dimension Styles and Variables dialogue box.

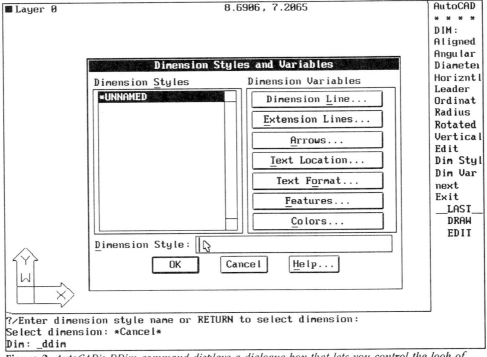

Figure 2: *AutoCAD's DDim command displays a dialogue box that lets you control the look of dimensions via style settings.*

Dimension Line

Click on the **Dimension Line** button to display the child dialogue box for setting the parameters of dimension lines (see Figure 3).

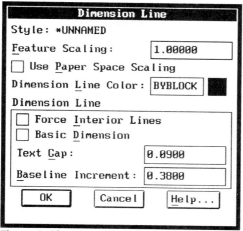

Figure 3: *The Dimension Line dialogue box controls the style of dimension line.*

- **Feature Scaling:** sets the scale factor for all dimensions (stored in the DimScale system variable).
- **Use Paper Space Scaling:** AutoCAD calculates the dimension scale based on the paper-space to model-space scale factor (DimScale system variable is set to 0.0).
- **Dimension Line Color:** the dimension line, extension line, text, and arrowhead can have a different color (stored in DimClrD system variable). When it comes time to plot, the dimension colors are mapped to pen thickness or pen color.
- **Force Interior Lines:** tells AutoCAD to draw dimension lines between the extension lines, even if text is forced outside due to the narrow distance between extension lines (system variable DimToFL).
- **Basic Dimension:** AutoCAD draws a rectangle around the dimension text.
- **Text Gap:** the length of the text gap in the dimension line (stored in DimGap).
- **Baseline Increment:** distance to offset baseline and continuous styles of linear dimensioning (stored in DimDLi variable).

Click on the **Cancel** button, then click on the **Extension Lines** button.

Extension Lines

The Extension Lines dialogue box lets you control the parameters AutoCAD uses to draw the extension lines (see Figure 4). It is similar to the Dimension Line dialogue box; the differences are discussed here:

- **Extension Above Line:** the distance that the extension line extends beyond the dimension line (stored in the DimExe system variable).
- **Feature Offset:** the gap between the extension line and the object being

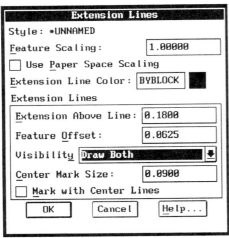

measured (stored in the DimExO variable).

- **Visibility:** you have the option of suppressing either or both extension lines (stored in DimSe1 and DimSe2 variables).
- **Center Mark Size:** the size of the center mark when dimensioning arcs and circles (stored in DimCen).
- **Mark with Center Lines:** toggles whether the center mark is drawn with center lines (a negative value in DimCen).

Figure 4: *The Extension Lines dialogue box controls the style of dimension extension lines.*

Click on the **Cancel** button, then click on the **Arrows** button.

Arrowheads

The Arrows dialogue box controls the parameters AutoCAD uses to draw the arrow heads (see Figure 5). Arrowheads are drawn in the same color as the dimension line.

- **Arrows:** you have a choice of using an arrowhead, a tick mark, a round dot, or specifying your own style of arrowhead.
- **Arrow Size:** this is the actual (not scale) length of the arrowhead (stored in DimASz).
- **User Arrow:** the user-defined arrow is simply an AutoCAD block, whose name you specify here (stored in DimBlk variable).
- **Separate Arrows; First Arrow; Second Arrow:** lets you specify a different arrowhead for each end of the dimension line.

Figure 5: *The Arrows dialogue box controls the style of dimension arrowheads.*

- **Tick Extension:** when using ticks for arrowheads, this is the distance the tick extends past the extension line (stored in DimDLE).

Click on the **Cancel** button, then click on the **Text Location** button.

Dimension Text

The Text Location and Text Format dialogue boxes control the parameters AutoCAD uses to draw the dimension text (see Figure 6). Dimension text is styled independent from other text in AutoCAD.

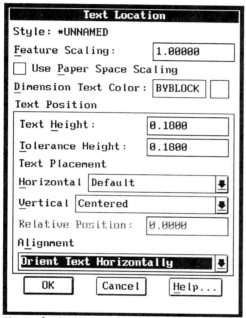

Figure 6: *The Text Location dialogue box controls the location of dimension text.*

- **Text Height:** the height of the characters in the dimension text (stored in the DimTxt system variable).
- **Tolerance Height:** the height of tolerance text (stored in the DimTFac variable).
- **Text Placement:** here you have two options:
  (1) horizontal placement of dimension text where it fits, forced inside the extension lines, or just forcing the arrows inside the extension lines (stored in DimTix and DimSoXd variables);
  (2) vertical placement of text centered on the dimension line, above the dimension line, or at a relative value to the dimension line (stored in DimTaD).
- **Alignment:** dimension text can be forced horizontal or aligned with the dimension line (stored in the DimTih and DimToh system variables).

Click on the **Cancel** button, then click on the **Text Format** button

- **Length Scaling:** this is a global scale factor applied to numerical values in dimension text and can be used for metric-imperial conversion (stored in the DimLFac system variable; see Figure 7).
- **Round Off:** the value to which all dimensions are rounded (stored in the DimRnd variable).

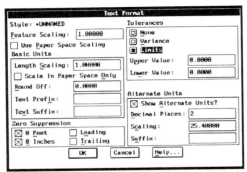

Figure 7: *The Text Format dialogue box controls the style of dimension text.*

- **Text Prefix; Text Suffix:** the prefix and suffix applied to all dimension text (stored in the DimPost variable).
- **Text Suppression:** toggles whether zero values are suppressed in dimension values (stored in the DimZin system variable).
- **Tolerances:** AutoCAD can generate tolerances (dimension text plus tolerance text) or limits (the dimension text is the tolerance text); stored in the DimTol, DimLim, DimTp, and DimTm system variables.
- **Alternate Units:** allows use of two measurement units in the dimension text, such as imperial and metric; the second unit is surrounded in parentheses (stored in the DimAlt, DimAltD, DimAltF, and DimAPost variables).

Click on the **Cancel** button, then click on the **Features** button.

All Dimension Features

The Features dialogue box combines the previous five dialogue boxes into one humongous box with the exception of specifying the colors (see Figure 8). The dialogue box nearly covers the entire screen of a VGA display. Use it when you need to set all parameters of the dimension style at once.

Click on the **Cancel** button, then click on the **Colors** button.

Dimension Element Colors

The Colors dialogue box combines the scaling and color options of the first three dialogue boxes. It lets you specify the color of the dimension line, extension line, and dimension text.

Scaling Dimensions

Like text in a drawing, dimensions cannot be drawn full size. Instead, you must apply scale factors to the size of the arrowhead and dimension text. The same calculations for scaling other text in the drawing applies to dimension; see Chapter 6, "Fonts, Linetypes, and Hatch Patterns."

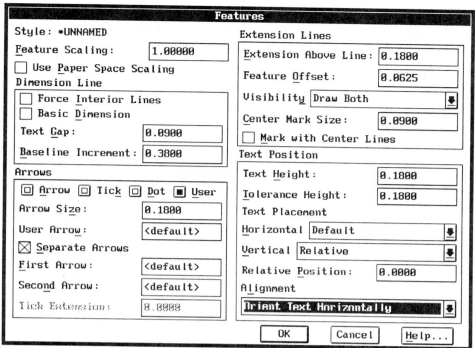

Figure 8: *The Features dialogue box combines all other dimension dialogue boxes with the exception of color parameters.*

Using Dimension Styles

If a CAD package, like AutoCAD, allows multiple dimension styles, then you need to be familiar with the method of saving and loading the styles.

AutoCAD has six commands for dimension style management. Before you can use them, you must enter dimensioning-command mode with the **Dim** command, as follows:

```
Command: dim
Dim: save
```

- **Dim: Save** saves the current setting of dimension variables as a style.
- **Dim: Restore** makes a dimension style current in the drawing.
- **Dim: Override** overrides the value of a dimension variable.
- **Dim: Update** updates all dimensions in the drawing with the current dimension style.
- **Dim: Variables** lists the values of the parameters associated with the current dimension style.
- **Dim: Status** lists the setting of all dimension-related system variables.

AutoCAD does not store dimension styles in a file on disk; styles are stored in the drawing. To move a dimension style from one drawing to another, use the **XBind Dimstyle** command to bind the style from an externally referenced drawing.

Summary

In this chapter you learned about the complexities of setting up dimension styles for your CAD system, in particular AutoCAD. In the next chapter, you put everything you've learned into a prototype drawing for eight different CAD packages.

CHAPTER 8

Preparing the Prototype Drawing

Now that you have decided on CAD drawing conventions for your firm, it is time to put them into a standard. There are two parts to creating an office standard: (1) embed the standards in a prototype drawing, and (2) write them up in a standards manual. This chapter describes how to prepare a standard prototype drawing for four major CAD packages and four low-cost CAD packages. The chapter following describes creating a standards manual.

What is a Prototype Drawing?

A prototype drawing eliminates the tedious work of the CAD operator setting up each new drawing to your firm's standards. A prototype drawing is very much like a spreadsheet template file and a word processor boilerplate file. The purpose of the prototype drawing is to eliminate repetitive work.

The prototype drawing acts as a template for all new drawings. It is a catalyst: it creates new drawings but is not consumed in the process.

A prototype drawing contains many predefined standard elements: layer names, colors, text styles, linetypes, dimension styles, and externally referenced drawings.

If your firm creates just one kind of drawing, you need create just one prototype drawing. If your firm is multi-disciplinary, create several prototype files—one for each kind of drawing your firm works with.

> ## Executive Summary
>
> *To create a basic prototype drawing, follow these 15 steps:*
>
> 1. Open a new drawing.
> 2. Select a unit of measurement.
> 3. Set the resolution of measurement.
> 4. Select a style and resolution of angular measurement.
> 5. Set the drawing's origin and extents.
> 6. Set object snap and drawing modes.
> 7. Load text fonts and define text styles.
> 8. Load linetypes and set linetype scale.
> 9. Create and name layers.
> 10. Load hatch patterns and pattern scale.
> 11. Load symbol libraries and external references.
> 12. Set dimension variables and scale.
> 13. Select the default plotter and set its plot style.
> 14. Save the prototype drawing.
> 15. Make two back-ups.
>
> Step-by-step instructions are given for the following eight CAD packages: MicroStation 5, AutoCAD Release 12, CadKey 6, Cadvance 5 for Windows, Drafix CAD Windows 2, Generic CADD 6.1, DesignCAD 6, and AutoSketch for Windows. ■

Examples of different prototype drawings you might want to consider are:

- Imperial or metric dimensions
- Two-dimensional or three-dimensional drawings
- Architectural, structural, mechanical, landscape, NC path, or electrical drawings

Whether you decide on one master prototype drawing or numerous specialized prototype drawings depends in part on the CAD system you use. For example, you don't need separate 2D and 3D prototype drawings in AutoCAD but you do in MicroStation and Cadvance. Drafix CAD Windows, Generic CADD, DesignCAD, and AutoSketch don't draw in three dimensions, so a 3D prototype drawing is a non-issue for them.

Preparing the Prototype Drawing

There are fifteen steps to preparing a standard prototype drawing:

1. **Open a new drawing file.** It forms the basis of the prototype drawing.

2. **Set unit of measurement.** The possibilities for units depend on the CAD package's capabilities. Some examples include:
   - Imperial, metric, or unitless measurement
   - Feet and fractional inches: 1'-6 1/4"
   - Feet and decimal inches: 1' 6.25"
   - Fractional or decimal inches without feet: 18 1/4" or 18.25"
   - Centimeters or millimeters: 18.25 cm or 182.5 mm
   - Scientific units: 18.25E+01

3. **Set the resolution of measurement.** Depending on the CAD package, the capabilities include:
   - For fractional inches, set measurement resolution to the nearest 1/2", 1/4", 1/8", 1/16", et cetera
   - For decimal inches, metric, scientific and unitless, set measurement resolution of decimal places, such as 1 (18.3) and 2 (18.25).

4. **Specify the style of angular measurement.** Depending on the CAD package, the capabilities include:
   - Direction of 0 degrees: North, West, South, East, or some other angle in between
   - Direction of angle measurement: clockwise or counterclockwise
   - System of measurement: decimal degrees (90.00), degrees-minutes-seconds (90d 00m 00s), grad (100.00g), radian (1.57r), or surveyor's (N 90d0'0" E)
   - Decimal number of places

5. **Set the drawing origin and extents.**

6. **Set object snaps and drawing modes.** Drawing modes include the grid and snap size, and orthographic or isometric mode.

7. **Load text fonts and define text styles.**

8. **Load linetypes and set linetype scale.**

9. **Create layers.** Give each layer a name, color, linetype, and default visibility.

10. **Create and load hatch patterns.**

11. **Load or reference symbol libraries.** This includes the title block and drawing border; if necessary, specify externally referenced files.

12. **Set dimension variables.**

13. **Select default plotter and set plot styles.**

14. **Save the prototype drawing.** Store it on disk with a special filename in a specific subdirectory.

15. **Make two backups of the prototype drawing.** Store one copy on-site; store the other copy off-site.

Once the prototype drawing has been created, it is not cast in stone. You edit it anytime to modify its contents. Be aware, though, that drawings based on the older prototype drawing are not automatically updated to reflect the new prototype drawing.

As you become familiar with your own way of working with CAD, you may want to add other elements to the prototype drawing. Some examples include named views, system variables, standard notes, customized menus, pre-programmed macros and programming routines, and third-party applications.

The remainder of this chapter describes how to create and use a prototype drawing for eight CAD packages, listed in order of decreasing price:

- **MicroStation Version 5** from Intergraph Corp.
- **AutoCAD Release 12** from Autodesk, Inc.
- **CadKey Version 6** from CadKey, Inc.; uses a hardware lock
- **Cadvance 5 for Windows** from IsiCAD, Inc.
- **Drafix CAD Windows Version 2** from Foresight Resources
- **Generic CADD 6.1** from Autodesk Retail Products
- **DesignCAD 2D Version 6** from American Small Business Computers
- **AutoSketch for Windows** from Autodesk, Inc.

In the following sections, command names are shown in **BoldFace** while menu picks are separated by a vertical bar, as in **File | Open**. Named keys, such as [Alt] and [Ctrl], are shown in square brackets, while keys you press simultaneously are joined by a plus sign, as in [Ctrl]+C.

MicroStation

With the fullest support for prototype drawings of any CAD package, MicroStation provides 17 template drawings (called "seed files") for the following disciplines (see Figure 1):

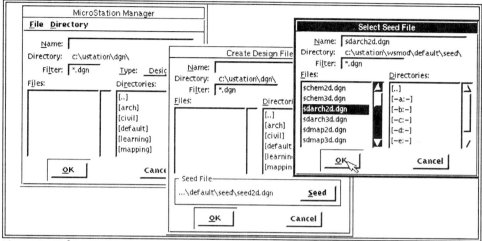

Figure 1: *Before starting a new drawing, MicroStation provides you with a list of seed files.*

- Generic 2D and 3D drawings
- Architectural 2D and 3D drawings
- Chemical 2D and 3D drawings
- Mapping 2D and 3D drawings
- Mechanical 2D and 3D drawings

Version 5 of MicroStation adds the concept of the "workspace," which customizes the user interface and keeps track of all project files. The CAD package includes the following workspaces found under the **User | Workspace | Select Default Workspace User** menu item:

- Architectural design
- AutoCAD-compatibility
- Civil engineering
- Mapping
- MDE programming
- Mechanical drafting
- Drawing translation
- ...and a simplified interface for new users

In addition, global settings are saved with the **File | Save Settings** command.

Open a New Drawing

Start MicroStation at the DOS prompt with the UStation.Bat batch file, as follows:

```
C:\> ustation
```

When you start MicroStation without a drawing, its MicroStation Manager dialogue box prompts you for a drawing name (see Figure 2).

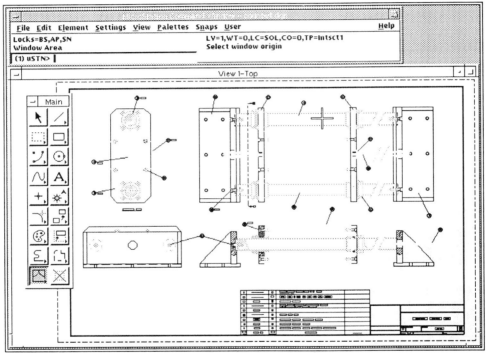

Figure 2: *MicroStation uses the OSF/Motif graphical user interface.*

Set Unit of Measurement

Initially, the measurement unit (called the "working unit") is set by the seed file. The **Settings | Working Units** command lets you set the name of the working units and resolution. MicroStation supports two kinds of units: master units (the units you work with in the drawing) and sub units (the working resolution).

MicroStation supports the following units:

- Feet-inches-thousandths
- Miles-feet-inches
- Millimeters-thousandths
- Yard-feet-inches

Set Resolution of Measurement

MicroStation is an integer-based CAD package that handles a maximum of 2,000,000,000 units in the x- and y-directions. Therefore, the resolution of measurement affects the size of the drawing. The disadvantage is that the finer the resolution, the smaller the drawing; the advantage is that you can pick any resolution you want. For example, selecting a resolution of 0.001" limits the drawing to about 117 square feet.

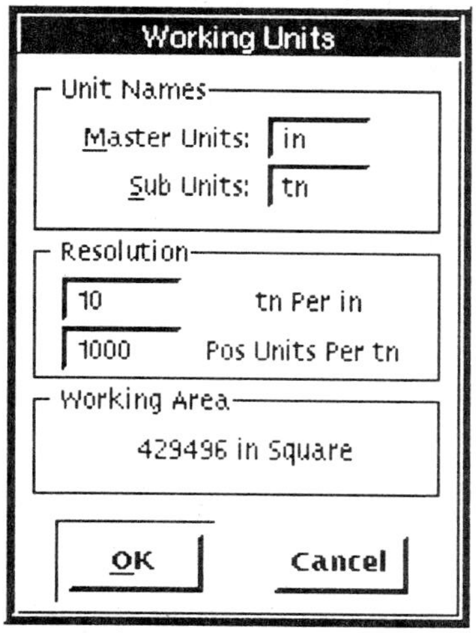

Thus, you set two values for the resolution: the smallest unit and the fractional (or decimal) accuracy. The accuracy is always larger than the smallest unit. MicroStation then calculates the world size (the largest dimension) and displays it at the bottom of the dialogue box. The **Settings | Coordinate Display** command lets you set the accuracy of display (see Figure 3):

- **Decimal:** zero to four decimal places
- **Fractional:** 1/2 to 1/64

Figure 3: *The Working Units dialogue box in MicroStation.*

Set Style of Angular Measurement

MicroStation has two forms of angle measurement, which are accessed by the **Settings | Coordinate Readout** command: decimal and degree-minutes-seconds.

Degrees are measured in three ways:
- **Conventional:** East is zero
- **Azimuth:** North is zero
- **Bearing:** relative to the quadrant

Accuracy of display is from zero to four decimal places.

Drawing Origin and Extents

MicroStation cannot relocate the origin of the drawing nor specify a limit to the extents.

Set Object Snaps and Drawing Modes

The 14 object snaps, found under the **Snap** menu, are (see Figure 4):

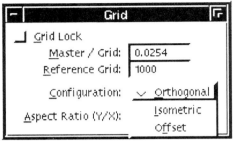

- Nearest
- Midpoint
- Center
- Origin
- Bisector
- Intersection
- Tangent and tangent point
- Perpendicular
- Parallel
- Through point

Figure 4: *The Grid dialogue box in MicroStation.*

In addition, MicroStation has a "tentative" point that moves the pick point between significant geometric features. The **Settings | Grid** command lets you set two grids: a larger, more obvious grid (called the "reference grid") and a second grid that is based on the master units. The x- and y-direction of the grid can be set independently. The dialogue box also lets you set an isometric grid. The **IS** command toggles the isoplane.

The **Settings | Locks | Full** command displays a dialogue box that lets you set and toggle the settings of 13 items, including the grid, snap, and isometric modes.

Load Text Fonts and Define Text Styles

The **Element | Text** command defines the text style (see Figure 5):

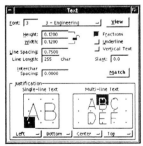

- Name of text font
- Height of characters
- Width of characters
- Spacing between characters
- Spacing between lines
- Slant angle
- Length of wrap
- Toggles for underlined, vertical, and fraction text
- Justification

Figure 5: *The Text dialogue box in MicroStation.*

The CAD package includes 18 fonts; a font conversion program (select **User | Utilities | Install Fonts**) reads and converts font definition files from AutoCAD SHX, PostScript PFB, and TrueType TTF font files.

Load Linetypes and Set Linetype Scale

MicroStation comes with 31 linetypes (called "line styles"). You create custom one- and two-dimensional line styles with the built-in Line Style Editor. The line style is set with the **Element | Line Style** command.

Create Layers

MicroStation has 63 layers (called "levels") numbered 1 to 63. You can assign a name and a DOS-like subdirectory-like structure and overcome the 63-layer limitation through reference files. MicroStation includes predefined level styles for the AIA layer standard (see Appendix C), divided by discipline to fit the 63-layer limit.

The **Settings | Level Symbology** command controls the color, line style, and line weight (width) of each layer (see Figure 6).

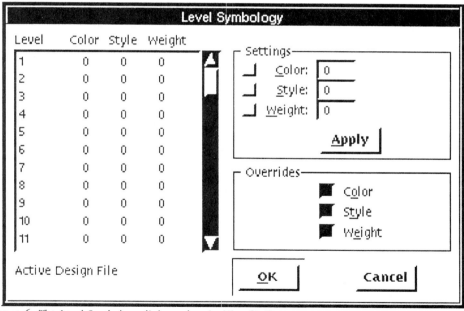

Figure 6: *The Level Symbology dialogue box in MicroStation.*

A second dialogue box, accessed with the **Settings | Level Names** command, lets you name levels, save level groups, and recall groups.

Load Hatch Patterns and Pattern Scale

MicroStation uses any symbol (or "cell") for placing hatch patterns (called "patterning"); "hatching" is drawing closely-spaced lines, which is defined on-the-fly. Since MicroStation uses cells for patterns, you can create custom patterns.

The **Palettes | Patterning** command displays the multi-purpose dialogue box that performs the following functions:

- Hatching
- Crosshatching
- Area patterning
- Linear patterning
- Display pattern attributes
- Match pattern attributes
- Erase a pattern

Load Symbol Libraries and External References

All MicroStation symbols (called "cells") are kept in CEL cell library files. The CAD package includes nine cell libraries.

Create cells with the **Settings | Cell | Create** command. While you can add 2D cells to a 3D cell library, 3D cells cannot be added to a 2D library.

Place cells with the same dialogue box, which features a cell previewer—it even previews 3D cells with a variety of shading algorithms! Before a cell can be used in a design file, its cell library must be attached with the **Settings | Cell | File | Attach** command.

MicroStation has very strong external reference capability, including the ability of drawings to reference themselves. Attach reference files with the **File | Reference** command.

Set Dimension Variables and Scale

MicroStation supports named dimension styles. Select a dimension style with the **Element | Dimensions | Placement | Style | Select** command:

- ANSI Y14.5 mechanical
- ISO mechanical
- DIN mechanical
- JIS mechanical
- Australian mechanical
- Intergraph INGR architectural
- Architectural graphics standards

Dimension units and accuracy are set independent of the design file via the **Element | Dimensions | Units** command. The dialogue box also lets you set a scale factor for the dimensions.

Select Default Plotter and Set Plot Style

Although MicroStation does not support multiple plot styles, its default values for each plotter is logical.

The Plot Preview feature is particularly powerful. It shows where the design file will be placed on the media and takes into account pen mapping to accurately predict the final outcome (see Figure 7).

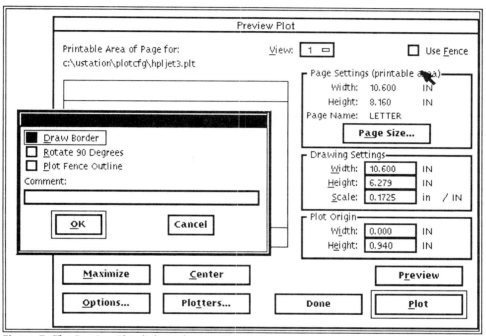

Figure 7: *The Preview Plot dialogue box in MicroStation.*

Save Prototype Drawing

With all the settings in place, save the drawing with the **File | Save As** command in the \Ustation\Wsmod\Default\Seed subdirectory. And don't forget to save the workspace with the **User | Workspace | Create Workspace User** command.

AutoCAD

When AutoCAD opens a new drawing, it uses either: (1) a prototype drawing named Acad.Dwg, (2) a prototype drawing name you specify, or (3) no prototype drawing at all. If your firm has need for only a single prototype drawing, then modify Acad.Dwg; otherwise, create a series of prototype drawings and load them implicitly (as described later).

Open a New Drawing

Start AutoCAD at the DOS prompt with the AcadR12.Bat file, as follows:

```
C:\> acadr12
```

When you start AutoCAD without a drawing, it automatically uses the Acad.Dwg prototype drawing to create a new, blank drawing named Untitled.Dwg (see Figure 8).

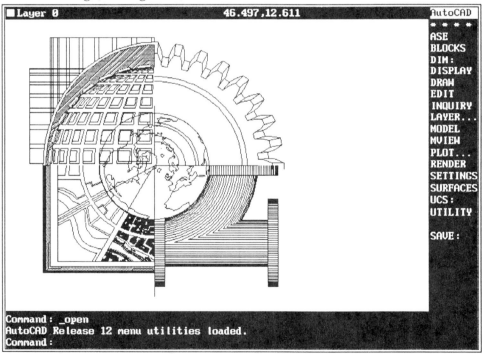

Figure 8: *The AutoCAD user interface combines its traditional look with a newer interface borrowed from Windows.*

To start a new drawing with a different prototype drawing (or with no prototype drawing at all), type the **New** command (or select **File | New** from the menu). The dialogue box gives you the following options (see Figure 9):

Figure 9: *The Create New Drawing dialogue box in AutoCAD.*

- Use Acad.Dwg as the default prototype drawing.
- Select a different drawing as the prototype
- Use no prototype drawing (required for a full DXF import)
- Retain the prototype drawing name as the new default

Set Unit of Measurement

A single command lets you specify the units and resolution of linear and angular measurement. Use the **DdUnits** command or select the **Settings | Units Control**. AutoCAD lets you set the following kinds of units (see Figure 10):

Figure 10: *The Units Control dialogue box in AutoCAD.*

- Scientific
- Decimal
- Engineering
- Architectural
- Fractional

The range is (in scientific notation) from 1E+308 (that's a one followed by 308 zeros) to 1E-308.

Set Resolution of Measurement

The same dialogue box lets you set the precision, as follows:

- Scientific: 0E+01 to 0.00 000 000E+01
- Decimal: 0 to 0.00 000 000
- Engineering: 0'-0" to 0'-0.00 000 000"
- Architectural: 0'-0" to 0"-0 1/256"
- Fractional: 0 to 0 1/256

The default is four decimal places (or 1/16") of accuracy. The architectural and engineering formats assume that one drawing unit equals one inch.

PREPARING THE PROTOTYPE DRAWING ■ 93

Set Style of Angular Measurement

The same dialogue box lets you select the style of measuring angles, as follows:

- Decimal degrees: D.ddd
- Degree-minute-second: DMS.ddd
- Grad: D.ddd g
- Radian: D.ddd r
- Surveyor: N DMS.ddd E

The precision ranges from zero to eight decimal places (or equivalent), as follows:

- Decimal degrees, grad, and radian: 0 to 0.00 000 000
- Degree-minute-second: 0d to 0d00'00.00 00"
- Surveyor: N 0d E to N 0d00'00.00 00" E

For the direction of 0 degrees, you choose from East (the default), North, West, South, or any point picked from the drawing. Angles are measured counterclockwise (the default) or clockwise.

Drawing Origin and Extents

The **Base** command changes the drawing's origin point. Use the **Ucs | Origin** command to move the user-coordinate system symbol from the screen's lower-left to the origin.

The **Limits** command defines the extent of the drawing, which limits the extents of the drawing grid and the Zoom All command.

Set Object Snaps and Drawing Modes

AutoCAD has ten object snap modes, as follows:

- Center
- Endpoint
- Insertion point
- Intersection
- Midpoint
- Nearest
- Node
- Perpendicular
- Quadrant
- Tangent

You set object snap modes before a command (using the **Osnap** command) or during a command.

The **DDrModes** command (or select **Settings | Drawing Aids**) displays a dialogue box to select drawing modes (see Figure 11):

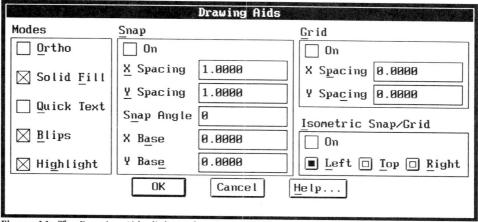

Figure 11: *The Drawing Aids dialogue box in AutoCAD.*

- Toggle orthographic mode or press [Ctrl]-O
- Set snap and grid spacing
- Toggle snap and grid modes or press [Ctrl]-B and [Ctrl]-G, respectively
- Toggle isometric mode and set the default isoplane: left, top, or right
- Snap angle defines the rotation of orthographic mode
- The curiously-named Snap Base defines where hatch patterns originate and the base point for rotated orthographic mode.

The x- and y-values for snap and grid spacing can be set independently.

Load Text Fonts and Define Text Styles

AutoCAD allows as many text styles in a drawing as you care to create. A point of confusion is that you cannot use text in an AutoCAD drawing until you first create a style. A single text font (such as Simplex) is commonly used for a number of styles. A single text style, called Standard (and uses the Txt font), exists in the Acad.Dwg prototype drawing.

The **Style** command creates a new text style, or select **Draw | Text | Set Style** from the menu. The Style command lets you set the following parameters:

- Text font file name
- Height of text
- Width factor of characters
- Obliquing angle of characters
- Draw text backwards
- Draw text upside down
- Draw text vertically

You set color and layer with the Color and Layer commands. A text height of zero (0.0 or 0") has special meaning in AutoCAD. Rather, than drawing text with no height, the Text command prompts you for a height.

AutoCAD comes with 38 fonts, each in an individual SHX font file. The Style command reads SHX and PFB PostScript Type 1 font files. You can create custom text fonts. Be aware that AutoCAD has a strange quirk that requires the SHP source code file be compiled to an SHX font file with the **Compile** command before the font is used for the first time.

Load Linetypes and Set Linetype Scale

AutoCAD comes with 9 linetypes stored in Acad.Lin. Other than the continuous linetype, each is supplied in double and half scale, for a total of 25 linetypes. You create custom linetypes in three ways:

- by adding new definitions to the Acad.Lin files
- create a new LIN file
- on-the-fly with the **Linetype Create** command

AutoCAD is limited to one-dimensional linetypes.

Before you use a linetype in a drawing, you must load it from Acad.Lin into the drawing with the **Linetype** command (there is no menu pick or dialogue box for this step) with the **Load** option. When prompted the linetypes to load, type an asterisk to load all linetypes, as follows:

```
Linetype(s) to load: *
?/Create/Load/Set: [Enter]
```

Set the linetype scale with the **LtScale** command.

Create Layers

AutoCAD is one of the few CAD systems that allows unlimited named layers in a drawing—up to 31 characters long. However, the CAD package does not support layer styles nor are any layer standards included. All AutoCAD drawings contain at least layer 0.

You create a new layer by naming it with the **DdLModes** command (or select **Settings | Layer Control** from the menu). The complicated looking dialogue box lets you perform the following functions (see Figure 12):

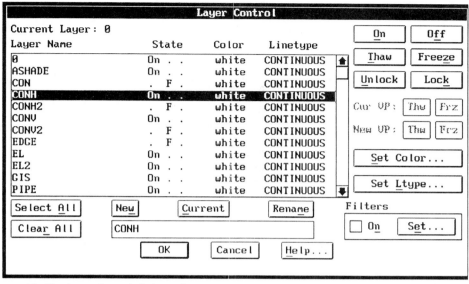

Figure 12: *The Layer Control dialogue box in AutoCAD.*

- Create a new layer by naming it
- Rename a layer
- Toggle layers: on and off; locked and unlocked; frozen and thawed
- Set up layer visibility for paper space
- Set color
- Select a linetype

You cannot associate a line width with a layer. An AutoCAD layer has four states:

- **Current:** objects added to the drawing are drawn on the current layer.
- **Off:** the objects are not displayed or plotted
- **Frozen:** similar to off, plus AutoCAD ignores objects during regeneration
- **Locked:** objects are displayed and plotted but cannot be edited

The **Select All**, **Clear All**, and **Filters** buttons allow you to create a selection set of layers displayed by this dialogue box.

Load Hatch Patterns and Pattern Scale

AutoCAD comes with 53 hatch patterns defined in the Acad.Pat file. You create custom hatch patterns by three ways:

- Add new definitions to the Acad.Pat file
- Place a single definition in another PAT file
- Define a hatch pattern on-the-fly with the **Hatch U** command.

The **BHatch** command (short for boundary hatch; or select **Draw | Hatch** from the menu) has the following options (see Figure 13):

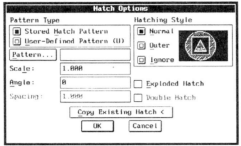

Figure 13: *The Hatch Options dialogue box in AutoCAD.*

- Use a predefined hatch pattern or make one on-the-fly ("user-defined pattern")
- Select the hatch pattern from the Acad.Pat file or other file
- Scale
- Angle
- Spacing between pattern lines
- Hatching method (called "style"): normal, outer, and ignore

- Draw the pattern as individual lines (called "exploded), rather than place as a symbol (called a "block")
  - Define the boundary and ray casting

The origin of the hatch pattern is hidden in the Snap command. If you just want to draw the hatch boundary, use the **BPoly** command.

Load Symbol Libraries and External References

AutoCAD does not support the concept of symbol libraries (called "blocks"). Blocks are stored in the drawing (with a 31-character name) or on disk in DWG files (with an eight-character name).

You create blocks with the **Block** command (or select **Construct | Block** from the menu). Save the block to disk with the **WBlock** command.

You place a block in the drawing with the **Insert** command. AutoCAD automatically searches the drawing for blocks and the DOS Path for DWG files. AutoCAD lacks a block preview feature.

To reference external drawings, use the **Xref** command; to "bind" (merge) an externally-referenced drawing to the current drawing, use the **XBind** command.

Set Dimension Variables and Scale

AutoCAD supports an unlimited number of dimension styles in a drawing; however, the CAD package does not include any dimension standards. The Acad.Dwg prototype drawing contains a single dimension style named *Unnamed.

To customize a dimension style, use the **DDim** command (or select **Settings | Dimension Style** from the menu).

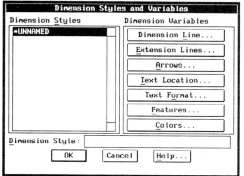

Figure 14: *The Dimension Styles and Variables dialogue box in AutoCAD.*

The Dimension Styles and Variables dialogue box (see Figure 14) lets you specify the following parameters (see Chapter 7, "Setting Up Dimensions," for more details):

- Dimension lines
- Extension lines
- Arrow heads
- Location of text
- Format of text
- Colors of individual parts

The Features child dialogue box summarizes many of the other dialogue boxes.

Select Default Plotter and Set Plot Style

AutoCAD lets you specify styles for 29 plotters; you can create multiple styles for a single plotter. AutoCAD has plotter drivers for vector, raster, PostScript, and common raster formats, such as TIFF, GIF, and PCX.

Creating a plot takes two steps: (1) configure AutoCAD for one or more plotters with the **Config** command, and (2) use the **Plot** command to create styles and plot the drawing.

The **Plot** command displays the dialogue box shown in Figure 15, on the next page.

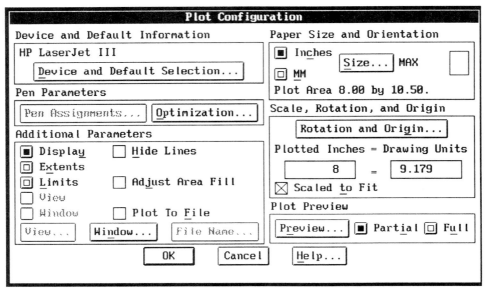

Figure 15: *The Plot Configuration dialogue box in AutoCAD.*

Save Prototype Drawing

Save the prototype drawing with the **SaveAs** command (or select **File | Save As** from the menu). When starting a new drawing, specify this as the prototype.

CadKey

There is no facility for prototype drawings in CadKey. The work-around is to merge an existing drawing into a new or current drawing.

CadKey has a Drawing Layout mode for adding details to the finished model, positioning views, and dimensioning in a 2D layout. CadKey's model and layout mode are similar to AutoCAD's model and paper space.

Open a New Drawing

Start CadKey at the DOS prompt with the CadKey.Exe program file, as follows:

```
C:\> cadkey
```

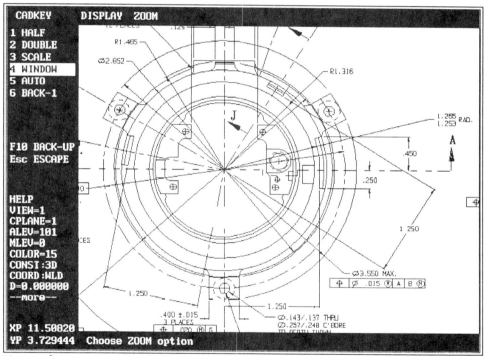

Figure 16: *For version 6, CadKey received a new dialogue box user interface.*

When you start CadKey without a drawing, it automatically creates a new, blank drawing with no name (see Figure 16). In CadKey, a drawing is known as a "part." Part filenames have the PRT extension unless otherwise defined by the user.

Set Unit of Measurement

Units are set in the Config program, which you must run before loading CadKey for the first time. Called "construction units," option #5, **Set Program Options**, lets you choose from:

- **Imperial:** inches, feet, or yards
- **Metric:** millimeters, centimeters, or meters

You change construction units from within CadKey using **Detail | Set**.

Set Resolution of Measurement

The Config program prompts you to set the decimal precision range between four and ten decimal places for CADL (CadKey advanced design language). In CadKey, the **Detail | Set** command sets dimension precision between zero and six decimal places.

Set Style of Angular Measurement

CadKey does not allow you to change the style of angle measurement. CadKey measures angles counterclockwise, unless you specify otherwise.

Drawing Origin and Extents

CadKey does not allow you to move the drawing's origin. The two work-arounds are (1) to move all objects in the drawing, and (2) use the **Display | Grd/Snap | Snap Aln** command.

Neither does CadKey have a limit the user can impose on the drawing. Instead, CadKey imposes a limit on entities themselves so that you don't get, for example, infinitely large arcs.

Set Object Snaps and Drawing Modes

CadKey has object snaps (called "positioning") under the **Position** menu:

- Cursor
- Point
- EndEnt (endpoint)
- Center of circular elements; midpoint of linear elements
- Intrsc (intersection)
- AlongL (distance along a line)

When you turn on an object snap in CadKey, it remains turned on until explicitly turned off. Snap and grid options are found under the **Display | Grd/Snp** menu:

- **Grid Dsp** (or press [Ctrl]-G) toggles the display of the grid
- **Grid Aln** repositions the grid
- **Grid Inc** sets independent grid spacing in the x- and y-directions
- **Grid=Snp** sets the grid increment to the snap increment
- **Snap Opt** (or press [Ctrl]-X) to toggle the snap
- **Snap Inc** sets independent snap spacing for the x- and y-directions
- **Snap Aln** repositions the snap
- **Snap=Grd** set the snap increment equal to the grid increment

CadKey does not have orthographic or isometric drawing modes.

Load Text Fonts and Define Text Styles

To set a text style before entering text (called "notes), use the **Set | Text** command. CadKey lets you specify the these parameters (see Figure 17):

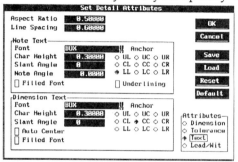

Figure 17: *Set Detail Attributes dialogue box.*

- Aspect ratio of characters
- Spacing between lines
- Name of text font
- Height of characters
- Slant angle of characters
- Slant angle of text lines
- Toggle font filling
- Toggle underlining
- Justification (called "anchoring")

CadKey includes four text fonts stored in individual FNT files; you can create custom fonts. Place notes with the **Detail | Note | Keyin** command or read a text file from disk. After placement, change text with the **Detail | Change** command or with built-in editor.

Load Linetypes and Set Linetype Scale

You set object attributes, such as line type, color, and line width with the **Control | Attrib** command. CadKey has four hardcoded linetypes, which are accessed by the [Alt]-T shortcut keystroke (called "immediate").

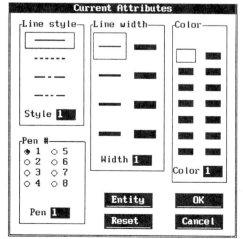

Figure 18: *Current Attributes dialogue box.*

Create Layers

CadKey has 256 numbered layers (called "levels") numbered from 1 to 256. You can assign a name of up to 30 characters to each level. CadKey has a form of level style via the **Display | Levels | Name | Txt-In** command. The CAD package does not supply any layer standards.

The **Display | Levels** command performs the following (see Figure 19):

- **Active:** sets the current layer to which objects are added
- **Remove:** turns off the display of levels
- **Add:** turns on the display of removed levels
- **Move:** moves objects from one level to another
- **Mask:** turns off all levels except those specified
- **Name:** names levels up to 30 characters
- **Entities:** counts number of entities on the level

The **Display | Levels | Name | List** command displays the following dialogue box summarizing the levels in the drawings, but only if at least one level has been named.

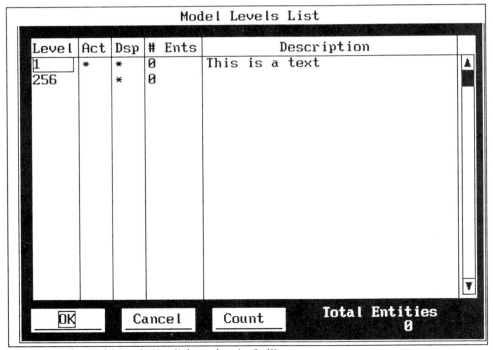

Figure 19: *The Model Levels List dialogue box in CadKey.*

Load Hatch Patterns and Pattern Scale

CadKey comes with 18 hatch patterns (called "cross-hatch styles"). You create custom cross-hatch styles in the file XHatch.Dat; you cannot create a hatch pattern on-the-fly. Cross-hatches in CadKey can be made associative with objects; you can cross-hatch 2D and 3D objects.

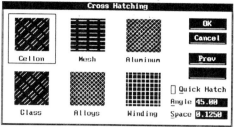

Figure 20: *The Cross Hatching dialogue box in CadKey.*

The **Detail | X-Hatch | Create** command displays the following dialogue box (see Figure 20):

- Select hatch style
- Specify the hatching angle
- Indicate the spacing between lines, which can be considered a form of scale factor

The Quick Hatch option lets you select the hatch area with a single cursor pick.

Load Symbol Libraries and External References

Create symbols (called "patterns") with the **Files | Pattern | Create** command. The pattern is saved to disk with a PTN extension. CadKey does not have symbol libraries.

To place a pattern in the part, use the **Files | Pattern | Retriev** command. The retrieved pattern is placed as a "group" or exploded. During placement, you can specify a different level number for every object in the pattern. CadKey does not support externally-referenced drawings.

The CAD package includes a set of ANSI-standard imperial and metric title block patterns:

- **Imperial:** A, B, C, D, and E
- **Metric:** A4, A3, A2, A1, and A0

Set Dimension Variables and Scale

CadKey supports dimension styles, called "templates," stored in DIM files; the CAD package does not include any predefined dimension styles. The **Detail | Set** command display a dialogue box for setting the dimension template (see Figure 21):

- Units of measure: inches, feet, yards, millimeters, centimeters, meters, or user-defined

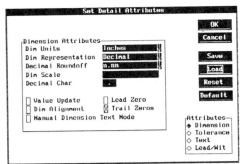

Figure 21: *The Set Detail Attributes dialogue box in CadKey.*

- Unit representation: decimal, fractional, feet-inch, or degree-minute-second
- Decimal round-off
- Scale factor for dimensions
- Replacement for the decimal point (such as the comma for European drawings)
- Value update (toggle associativity)
- Alignment: vertical or horizontal to the leader
- Toggle manual or automatic dimensioning
- Toggle leading and trailing zeros
- Tolerances
- Dimensions text attributes (separate from notes attributes)
- Leader and witness line attributes
- Arrowhead style

After you specify a dimension template, you change the attributes with the **Detail | Change** command.

Select Default Plotter and Set Plot Style

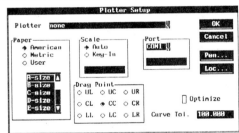

Figure 22: *The Plotter Setup dialogue box in CadKey.*

CadKey lets you plot from within the program (hidden in the **Control | Plot** menu pick) or outside with the PlotFast utility (see Figure 22). CadKey saves a single plot setup for all plotters, created with the Config program.

Save Prototype Drawing

While CadKey does not support prototype drawings, you can save the settings in drawings, as well as levels and dimension styles in separate files. Save the "prototype" drawing with the **Files | Part | Save** command.

Load the prototype with the **Files | Part | List/Ld** command, making a copy with the **Copy** button in the Part File Load dialogue box. Alternatively, write a macro to automate the copy process.

Cadvance for Windows

The concept of a "prototype" drawing does not exist in Cadvance. Instead, Cadvance uses configuration files that set up templates for the layer structure, reference drawings, dimension styles, and program configuration. Use the **Options | Path** command to point to a subdirectory.

Open a New Drawing

Start Cadvance under Windows by double-clicking on the Cadvance icon; a dialogue box prompts you for your initials—this lets Cadvance keep track of itself while running on a network (see Figure 23).

When you start Cadvance without a drawing, it creates a new, blank VWF drawing without a name (see Figure 24).

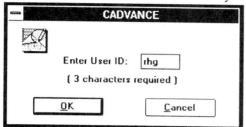

Figure 23: *Users must enter a three-letter identifier before entering Cadvance.*

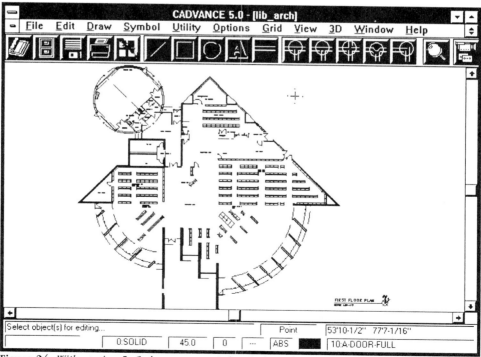

Figure 24: *With version 5, Cadvance converted entirely to the Windows operating environment.*

Set Unit of Measurement

To set units in a drawing, select **Options | Unit** from the menu (see Figure 25). The dialogue box lets you select a form of measurement:

- **Imperial:** feet-inch, inch, or foot
- **Metric:** millimeter, centimeter, or meter
- **Unitless:** unit

With the feet-inch units, you have the option of selecting one of three display formats: ft' in fr" or ft'-in fr" or ft' in-fr" (where ft = feet, in = inch, and fr = fraction).

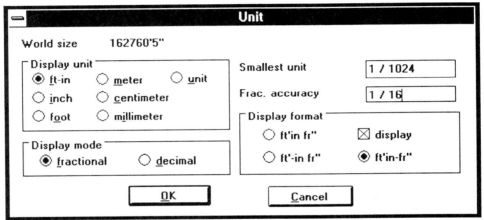

Figure 25: *The Units dialogue box in Cadvance.*

Set Resolution of Measurement

Cadvance is an integer-based CAD package that handles a maximum of 2,000,000,000 units in the x- and y-directions. Therefore, the resolution of measurement affects the size of the drawing. The disadvantage is that the finer the resolution, the smaller the drawing; the advantage is that you can pick any resolution you want. For example, if you select a resolution of 1/6", the largest dimension in the drawing is about 2,000 miles.

Thus, you set two values for the resolution: the smallest unit and the fractional (or decimal) accuracy. The accuracy is always larger than the smallest unit. Cadvance then calculates the world size (the largest dimension) and displays it in the upper left corner of the dialogue box.

Set Style of Angular Measurement

You cannot change the method and accuracy of angular measurement in Cadvance.

Drawing Origin and Extents

Cadvance does not allow you relocate the origin of the drawing; the workaround is to select all objects and move them.

The limits to the drawing are defined by the smallest unit, as described above. Extents are only defined for 3D drawings via the **Grid | Extents** command.

Set Object Snaps and Drawing Modes

Cadvance has five object snaps. They are accessed by clicking on seventh field of the two-line status area at the bottom of the screen or just press the period (.) key:

- Grid
- Node
- Vertex
- Intersection
- Line

While the list seems short compared to other CAD systems, these are fairly versatile. For example, the Vertex object snap mode snaps to the vertex point of any object, including quadrants of a circle. You have only one object snap turned on at a time.

Turn on the grid with **Grid | Display**; size the grid's x- and y-distance with **Grid | Size**. You can move the grid, as well as store and recall up to ten grid styles. Press the period (.) key, then the G key to snap to grid mode.

Cadvance does not have a snap mode as in other CAD packages. To use orthographic and isometric modes, you must switch to Cadvance to three dimensions by selecting **3D** on the menu bar. The **View** item's **Perspective**, **Axonometric**, and **Oblique** options give you access to orthographic and isometric modes.

Load Text Fonts and Define Text Styles

Cadvance includes 26 fonts defined in individual FNT font files; however, only five fonts can be loaded into a drawing at one time. Although you cannot create custom fonts, the newer Cadvance version 6 (not available at time of writing) reads TTF TrueType font files.

To choose the five fonts for the drawing, select **Options | Font**. To set a text style (Cadvance supports a single text style), the **Draw | Text** command displays the following dialogue box (see Figure 26):

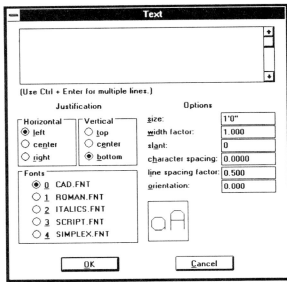

- Select the text font
- Justification
- Height of text
- Width factor of characters
- Slant angle of characters
- Slant angle of text line
- Spacing between characters
- Spacing between lines

Once text has been placed, it is edited with the **Edit | Text | Format** command.

Figure 26: *The Text dialogue box in Cadvance.*

Load Linetypes and Set Linetype Scale
Cadvance comes with eight linetypes (called "line styles"); three are hardwired: solid, dot, and dash. Up to eight are used in a drawing. You can create custom line styles by editing the Cad.Lns file.

Cadvance is more flexible than most CAD packages by allowing line "textures" that takes any symbol and turns it into a line style via the **Draw | Texture | Line** command. You cannot set the linetype scale.

Create Layers
Cadvance has 255 layers numbered from 1 to 255; a layer can be assigned a name. You create custom layer configurations in LYR files; the CAD package includes layer standards from the AIA for these disciplines:

- Architectural
- Civil engineering
- Electrical
- Fire protection
- Landscaping
- Mechanical
- Plumbing
- Structural
- Title blocks

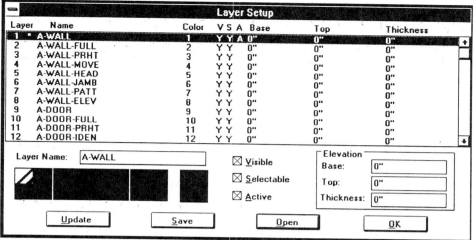

Figure 27: *The Layer Setup dialogue box in Cadvance.*

The Layer Setup dialogue box lets you set color, elevation, and thickness of each layer. Line styles are set independent of layers (see Figure 27). A layer in a Cadvance drawing has three states:

- **Active:** objects are added to the active layer
- **Visible:** objects are visible
- **Selectable:** objects can be edited

Load Hatch Patterns and Pattern Scale

Cadvance does not have any predefined hatch patterns (called "area textures") as other CAD packages provide. Instead, the **Draw | Texture | Area** command lets you create any area texture you want on-the-fly from any predefined symbol or just plain lines (see Figure 28, on the next page):

- Spacing: a form of scale factor
- Angle
- Spacing: even, random, or semi-random (creates illusion of curvature)

Cadvance has two kinds of texture fills, depending on whether you want to fill an existing object or an arbitrary area. *Area* texture fills existing objects. *Phantom* texture fills an area defined by your pick points in the drawing.

PREPARING THE PROTOTYPE DRAWING ■ 111

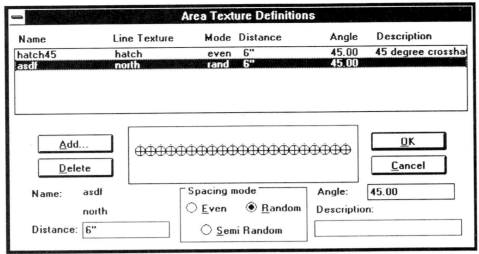

Figure 28: *Area Texture Definition dialogue box in Cadvance.*

Load Symbol Libraries and External References

The **Symbol | Make** command creates a new symbol and stores it in a SYM file on disk.

Place a symbol with **Symbol | Place**, which lets you select from symbols in the drawing or on disk.

To reference a drawing, use the **File | Open** command and click on the Read Only checkbox. Reference drawings can be saved as a REF file to prevent further changes and reference file setup information is saved in a RFS file. Loading an RFS file loads all referenced files.

Set Dimension Variables and Scale

Cadvance supports a single dimension style in a drawing. Multiple dimension styles are saved in Cad.Dim files. The **Draw | Dimension | Setup** command lets you specify the following parameters:

- Toggle automatic or manual dimensioning
- Text alignment, accuracy, and location in radial dimensions
- Type of center mark
- Orientation
- Style of extension
- Layer
- Color, line style, line weight
- Tolerance

Figure 29: *The Dimension Parameters Setup dialogue box in Cadvance.*

After placing the dimensions, you edit them with the **Edit | Properties** command.

Select Default Plotter and Set Plot Style

Cadvance uses the Windows system printers for output. The **File | Print Setup** command uses the dialogue box common to all other Windows applications. The **File | Print** command's dialogue box has few options: print window, margins, scale, and line weight (see figure 30). You can save a print configuration for each plotter.

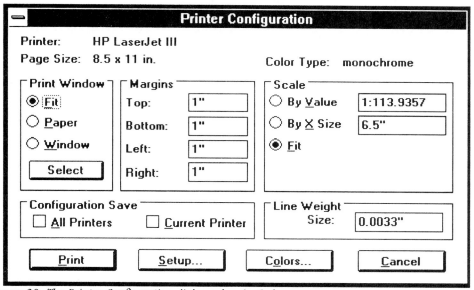

Figure 30: *The Printer Configuration dialogue box in Cadvance.*

Save Prototype Drawing

While Cadvance does not have prototype drawings, it allows you to save major parameters to files on disk:

- Layer configurations in *.Lyr
- Reference file configurations in *.Rfs
- Dimension configurations in *.Dim
- Program configuration in *.Ini

Each configuration file can be customized to your firm's line of work. In addition, you can make a drawing read-only, load it, then use the **File | Save As** command to immediately rename it.

Drafix Windows CAD

Drafix Windows CAD and other lower-priced CAD packages do not have an easy facility for using a prototype drawing. The workaround is to set parameters and save the drawing as a template in read-only mode.

Open a New Drawing

Start Drafix CAD under Windows by double-clicking on its icon. When you start Drafix CAD without a drawing, it automatically creates a new, blank drawing named Untitled.Cad (see Figure 31).

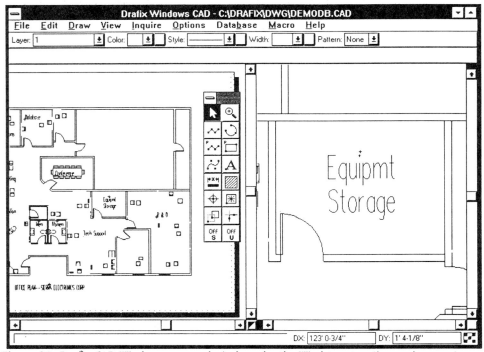

Figure 31: *Drafix CAD Windows runs exclusively under the Windows operating environment.*

Set Unit of Measurement

Set the measurement units by selecting the **Options | Units of Measurement** menu item. Drafix CAD displays a dialogue box with many options (see Figure 32). You select from the following units:

- **Imperial:** inches, feet, feet-inches, inches-fraction, feet-fraction, and feet-inches-fraction
- **Metric:** meters, centimeters, and millimeters

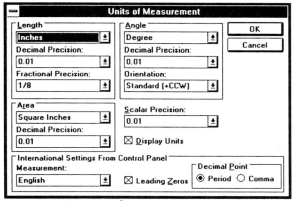

Figure 32: *The Units of Measurement dialogue box in Drafix CAD Windows.*

Unique among the CAD packages mentioned here is the ability of Drafix CAD to report area in units different from linear measurement. The units of measurement available in Drafix CAD are:

- **Imperial:** square inche, square foot, square yard and acre
- **Metric:** square millimeters, square centimeters, square meters, and hectares

In addition, Drafix CAD recognizes the international setting from the Windows Control Panel.

Set Resolution of Measurement

The same dialogue box lets you set the measurement resolution to the following ranges:

- **Fractional:** 1 to 1/128
- **Decimal:** 1.0 to 0.00 000 01

Drafix CAD has an independent setting for the resolution of scale factors, aspect ratios, and other unitless numbers. The scalar precision range is the same as decimal: 1.0 to 0.00 000 01.

Set Style of Angular Measurement

Drafix CAD reports angular measurement in the following methods:
- Degree: D.ddd
- Degree-minute-second: DMS.ddd
- Minutes: M.ddd
- Radian: D.ddd r
- Bearing: N D.ddd E D.ddd

The range of angular resolution is 1.0 to 0.00 000 01, while angles are measured clockwise (called "compass") or counterclockwise, called "standard." There is no means to change the direction of 0-degrees from East.

Drawing Origin and Extents

To move the drawing's origin and set the drawing limits, select **Options | Sheet Size** from the menu bar (see Figure 33). You specify a custom drawing size or select from standard sheet sizes:

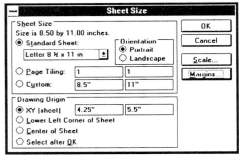

- **Imperial:** A, B, C, D, and E
- **Metric:** A4, A3, A2, A1, A0, B5, and C5
- **Other:** letter, legal, Monarch, executive, Com-10, and DL

The 18 sizes can be oriented in portrait or landscape. Drafix CAD draws a rectangular outline showing the drawing extents, complete with a drop shadow. You can draw outside the border.

Figure 33 *The Sheet Size dialogue box in Drafix CAD Windows.*

To set the drawing origin, you specify the x,y-coordinates, select the lower-left corner, the center, or pick a point on the drawing. A small double-arrow icon relocates to show the drawing origin.

Set Object Snaps and Drawing Modes

To access snap modes in Drafix CAD, click and hold on the bottom-left icon of the Toolbox. A tool set appears. While holding down the mouse button, drag across to select the object snap mode. Drafix CAD includes the following nine modes:

- Endpoint
- Midpoint
- On entity
- Intersection
- Perpendicular
- Center point
- Tangent
- Quadrant
- Grid point

To specify the grid, select **Options | Reference Grid**. You specify an independent x,y-grid made of dots and lines. To toggle the display of the grid, select **Options | Display**, then pick the check box next to Reference Grid.

To draw an orthogonal line, hold down the **H** or **V** key while drawing the line. This differs from the Windows standard of holding down the [Shift] key. Drafix CAD does not have isometric drawing aids.

Load Text Fonts and Define Text Styles

Defining a text style is probably the easiest in Drafix CAD. When you enter text mode (by clicking on the Toolbar's T-icon), the "CAD Edit Bar" (located under the status bar) changes to let you set:

- Text font name
- Height of text
- Width factor
- Aspect ratio
- Justification
- Angle of text line

The Status Bar lets you set the color and layer. The CAD package includes 19 fonts stored in a single file named Drafix.Fon. You cannot create custom fonts or import fonts from other sources.

Load Linetypes and Set Linetype Scale

Drafix CAD has seven hardwired linetypes (called a "pen style" or just "style," for short) named solid, short dash, long dash, center line, phantom, dotted, and dash dot. You select a line style from the Status Bar; you cannot create custom line styles.

Create Layers

Drafix CAD drawings hold up to 256 named layers. By default, layer 1 exists in every drawing. A layer name is up to 12 characters long; Drafix CAD allows you to add a 63-character description to each layer. To create new layers, select **Options | Layer Setup**. As shown by the dialogue box, a layer has four possible states (see Figure 34):

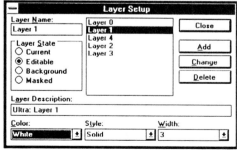

Figure 34: *The Layer Setup dialogue box in Drafix CAD Windows.*

- **Current**: the layer to which objects are added
- **Editable**: objects can be changed but not added

- **Background:** objects can be seen but not changed
- **Masked:** objects cannot be seen nor edited

To each layer, you set the following attributes:

- State (as described above)
- Color
- Line type (called "style")
- Line width

To switch to another layer, simply select the name from the drop box on the Status Bar.

Load Hatch Patterns and Pattern Scale

Drafix CAD hides its hatching parameters in the **Draw | Setup** dialogue box (see Figure 35). You select the style, spacing, and angle. "Style" is Drafix CAD name for hatch pattern name, while "spacing" means scale factor. The CAD package comes with 15 hardwired hatch patterns. You cannot create a custom hatch pattern, nor can you specify the hatch origin.

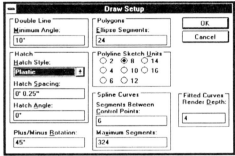

Once the spacing and angle are set in the dialogue box, you select the style from the Status Bar. In addition to hatch patterns, you can solid fill areas with colors.

Figure 35: *The Draw Setup dialogue box in Drafix CAD Windows.*

Load Symbol Libraries and External References

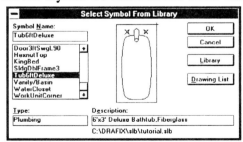

You create symbols with the **Edit | Symbol Definitions | Create** command. The dialogue box lets you specify the name, the location of the insertion point (called the "basepoint"), a pair of attribute fields (symbol type and description), and whether you want the symbol added to a library. To place a symbol, select **Draw | Symbol | Select**; Drafix

Figure 36: *The Select Symbol from Library dialogue box in Drafix CAD Windows.*

CAD lets you select the symbol from a list, from those already placed in the drawing, or those stored in a library. A handy preview feature lets you see the symbol before selecting it (see Figure 36).

The CAD package includes 450 symbols sorted into three SLB library files for the following disciplines:

- **Architectural:** building, landscape, and furnishings
- **Electrical:** electronic and PCB
- **Mechanical:** nuts, screws, threads, washers, welding, and structural.

Set Dimension Variables and Scale

To customize the look of dimensions, select **Draw | Dimension | ...** from the menu, where "..." refers to any dimension option from the child menu. Then select the **Format** button from the CAD Edit Bar.

Drafix CAD supports a single dimension style in a drawing. The dialogue box presents the following parameters (see Figure 37):

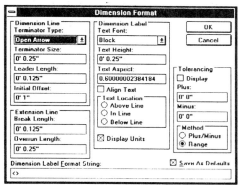

- Arrowhead (called a "terminator"): type and size
- Leader length
- Dimension line offset
- Extension line: break and overrun lengths
- Text font, height, aspect ration, justification, location, and label string
- Toggle display of units
- Tolerance: toggle display, plus, minus, and method

Figure 37: The Dimension Format dialogue box in Drafix CAD Windows.

The Status Bar lets you set the layer and color for the dimensions.

Select Default Plotter and Set Plot Style

As common with Windows-based CAD packages, Drafix CAD uses the Windows system printers. **File | Printer Setup** selects the printer, while **File | Print** outputs the drawing. Drafix CAD allows you to tile the drawing over multiple sheets of paper, a feature hidden in the **Options | Sheet Size** dialogue box.

Save Prototype Drawing

Drafix stores its parameters in the Drafix.Ini initialization file. This ASCII file can be edited to quickly set up parameters specific to a drawing type:

- **[Display]:** Display Options dialogue box
- **[Macro]:** Automatic start up macros and the DGL interpreter
- **[Path]:** Default directory paths
- **[Preferences]:** Preferences dialogue box
- **[Startup]:** Startup options
- **[Window]:** Desktop layout

By combining read-only template files with AutoExec macros, you create prototype drawings for Drafix CAD.

Generic CADD

The concept of multiple prototype drawings does not exist in Generic CADD. Instead, Generic CADD saves the current drawing environment to the Environ.Gcd and Environ.Fil settings file. There is no facility for multiple environments. Instead, place different Environ.* files in individual subdirectories, then use the **FP** command (short for file paths) to specify the path to the subdirectory.

Open a New Drawing

Start Generic CADD at the DOS prompt with the Cadd.Exe loader program, as follows:

```
C:\> cadd
```

When you start Generic CADD without a drawing, it automatically creates a new, blank drawing named Untitled.Gcd (see figure 38).

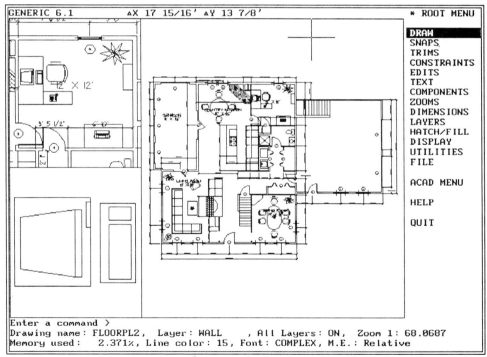

Figure 38: *Generic CADD's user interface is similar to AutoCAD's and version 6.1 even includes an AutoCAD-compatible menu structure.*

Set Unit of Measurement

Initially, the units of measurement are set in the Config.Exe configuration program. Option 4, Set CADD Database Unit, lets you set:

- **Imperial:** inches and feet
- **Metric:** millimeters, centimeters, and meters.

You can change the units within Generic CADD via the **Display | Units** command. However, Autodesk Retail Products warns that there may be some round-off error when switching between imperial and metric units.

Set Resolution of Measurement

While the **UN** command selects the measurement unit, the **NF** command (short for numeric format; or select **Display | Num Display** from the menu) controls the format, as follows:

- **Linear format:** decimal or fraction
- **Decimal value:** 0 to 6 decimal places
- **Fractional value:** 1/2 to 1/64th
- **Leading zeros:** on or off
- **Display units:** toggles the display of units on the status line

Set Style of Angular Measurement

The **NF** command also toggles the angular format between:

- **D.ddd:** degrees and decimal degrees
- **DMS:** degrees, minutes, and seconds

The command's decimal value option affects the display of linear and angular decimal places.

Drawing Origin and Extents

You specify the origin of a drawing when you first load it. That lets you insert multiple drawings, each with a different origin. Later, you change the origin with the **Utilities | Dwg Origin** command.

To set the drawing's extents, use the **Utilities | Limits** command.

Set Object Snaps and Drawing Modes

Generic CADD has its object snaps under the **Snaps** menu item. The 13 object snap modes are:

- Center
- Closest point
- Intersection
- Midpoint
- Nearest point
- Object
- Parallel
- Percentage
- Perpendicular
- Tangent
- Tolerance
- Component
- Quick pick

Object snaps do not stay on. You must respecify a snap mode each time you want to use it. Thus, there is no command to turn off an object snap mode.

Turn on the grid with the **Constraint | Grid On/Off** command; set the size with the **Constraint | Grid Size** command. You can specify different grid size in the x- and y-directions, as well as move the grid origin with **Constraint | Grid Reorigin**. To snap to the grid, use the **SG** command or press [F1].

Orthographic mode is turned on with **Constraint | Ortho Mode** or by pressing [F3]. You specify an angle for ortho mode with **Constraint | Ortho Angle**. Generic CADD does not have an isometric drawing mode.

Load Text Fonts and Define Text Styles

Generic CADD allows only a single text style at a time in a drawing, set with the **TS** command or **Text | Txt Settings** menu selection. The workaround is to use the **Text | Match Params** command to set the current text style based on an style used earlier in the drawing.

The TS command loads a text font from disk and sets the style. Generic CADD lets you specify the following options:

- Font name
- Text height
- Fill toggle
- Justification

- Angle of slant of text characters
- Rotation angle of text line
- Color of text
- Spacing between characters, lines; proportional toggle

Except for fill mode and spacing, the settings do not apply text in dimensions; these are set with the **US** command.

Generic CADD v6.1 comes with 37 text fonts, each located in an *.Fnt font definition file. A conversion utility, Gcd2Dwg.Exe, converts AutoCAD's SHX compiled text font files into Generic CADD's FNT format.

Load Linetypes and Set Linetype Scale

Generic CADD has ten hardwired linetypes numbered 0 to 9; you do not load linetypes from disk nor can you create custom linetypes nor can you create a linetype on the file. Linetypes are strictly one-dimensional.

You set the current linetype and its scale with the **LT** command (short for linetype) or the **Display | Line Type** menu pick. Linetypes affect the following entities:

- Line
- Arc, circle, curve, and ellipse
- Text
- Hatching

Linetypes 0 to 9 are scale independent. Linetypes 10 through 250 increase the scale factor. The **LZ** command changes the linetype scale factor.

Create Layers

Generic CADD has 256 layers numbered 0 to 255. You can assign an 8-character name to each layer with the **YN** command (short for laYer Name) or pick the **Layers | Name** command.

Turn off the display of a layer with **YH** or **Layers | Hide** command; turn on the display with the **YD** or **Layers | Display** command. Layers that are not displayed cannot be edited, changed, or plotted.

The **YG** or **Layers | Change** lets you set the following attributes for a each layer (but is limited to changing one layer at a time):

- Color
- Linetype
- Line width

Once you have defined the layers in a drawing, you can save selected layers (along with their contents) to a GCD file with the **YS** command. Unnamed layers are saved with the generic name of Layer001. Load the layers back into the drawing **YL** command; this command also loads a drawing from disk onto a single layer in the current drawing.

Load Hatch Patterns and Pattern Scale

Generic CADD comes with 50 hatch patterns defined in individual *.Hch files. You define custom hatch patterns by creating a new HCH file. In addition to patterns, Generic CADD can fill an area with a solid color. You cannot create a hatch pattern on the fly.

As with text fonts, a single command loads the hatch pattern and sets its parameters. The **HS** command (short for Hatch Settings; or **Hatch/Fill | Hch Settings** menu pick) has the following options:

- Set current hatch pattern by name
- Scale
- Rotation angle
- Color
- Display toggle
- Boundary toggle

Load Symbol Libraries and External References

Create symbols (called "components" by Generic CADD) with the **CC** command (short for create component) or **Component | Create** menu pick.

Once a symbol is created, you save it to disk in two ways:

1. **SA C** command (save component) saves individual components to a CMP file on disk.

2. **CD** command (component dump) saves all components in the drawing to individual CMP files on disk.

In both cases, the filename is either the component's name or you can specify a different name.

Load a component from disk with the **LO C** (load component) command, then place the symbol in the drawing with the **CP** command.

Generic CADD does not reference external drawings; the work-around is to load the second drawing onto a single layer with the **YL** command.

Set Dimension Variables and Scale

Generic CADD supports a single dimension style in a drawing. The **US** command (or pick **Dimensions | Dim Settings** from the menu) lets you specify the following parameters:

- Dimension text
- Text font
- Tolerance text
- Extension lines
- Arrow head size and type
- Dimension line
- Layer
- Color

After the dimension style is set, use the **UG** (or **Dimensions | Dim Change**) command to change the look of the dimension.

Select Default Plotter and Set Plot Style

Generic CADD lets you specify devices for three kinds of plotters: vector plotter (such as a pen plotter), raster plotter (such as a dot-matrix printer), and PostScript printer. The **File** menu lets you select **Plot**, **Print**, and **Postscript**, which brings up a text screen (see Figure 39):

```
                ******    GENERIC CADD    ******
                          Version  6.1

      Output menu for: H.P. 7475A PLOTTER

        1) Send to (plotter, printer, postscript): Plotter
        2) Port (COM1, COM2, LPT1, LPT2, LPT3, File): COM2
        3) Page size: (Length: 10.150, Width: 7.800)"
        4) Options (Configure Plotter, Pens, etc...)
        5) Select view type: Fit full drawing
        6) Page setup
        7) Start plot

   ESC)   Return to drawing


 Enter selection >
```

Figure 39: *The Plot command's menu screen in Generic CADD.*

Use Option 4, Options: Configure Plotter, to set the default plotter. You cannot save more than one style per kind of output device.

Save the Prototype Drawing

With all settings complete, save the drawing's environment with the **EN** or **File | Environment** command. While Generic CADD does not support multiple prototype drawings, you can save different environments to individual subdirectories.

As an alternative, save the drawing with the **SA** command. Load the template drawing with the **LO** command, then rename the drawing with the **DN** command (**File | Dwg Rename**). You can also change the name when you save the drawing.

DesignCAD 2D

Like other low-cost CAD packages, DesignCAD does not support prototype drawings. DesignCAD stores parameters in each drawing with the [Alt]-Q keystroke, or select **Options | Misc Options** from the menu.

Open a New Drawing

Start DesignCAD at the DOS prompt with the Dc.Exe program, as follows:

```
C:\> dc
```

When DesignCAD loads without a drawing, the empty drawing is unnamed (see Figure 40).

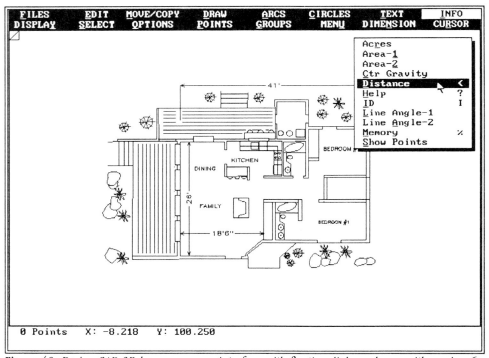

Figure 40: *Design CAD 2D has a new user interface with floating dialogue boxes with version 6.*

Set Unit of Measurement

The **U** command (or select **Options | Units** from the menu) lets you define an arbitrary set of units: use any system measurement you want. Pick two points on the screen, then enter the number of units it represents. Naturally, the recommended distance is one unit (see Figure 41).

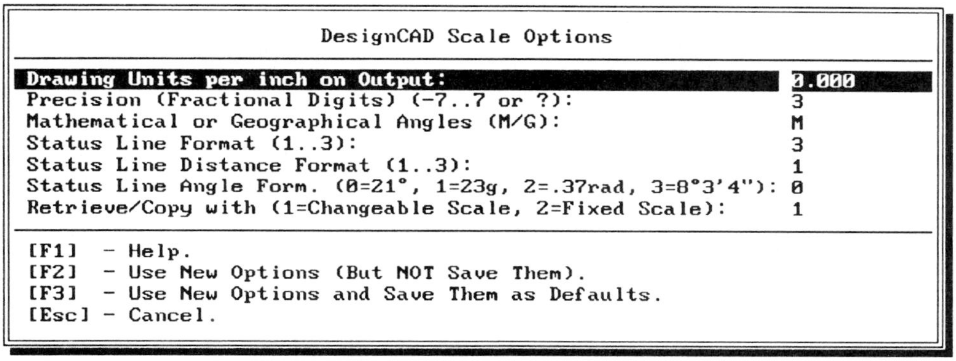

```
                    DesignCAD Scale Options
┌─────────────────────────────────────────────────────────────────┐
│ Drawing Units per inch on Output:                      0.000    │
│ Precision (Fractional Digits) (-7..7 or ?):            3        │
│ Mathematical or Geographical Angles (M/G):             M        │
│ Status Line Format (1..3):                             3        │
│ Status Line Distance Format (1..3):                    1        │
│ Status Line Angle Form. (0=21°, 1=23g, 2=.37rad, 3=8°3'4"): 0   │
│ Retrieve/Copy with (1=Changeable Scale, 2=Fixed Scale):  1      │
├─────────────────────────────────────────────────────────────────┤
│ [F1]  - Help.                                                   │
│ [F2]  - Use New Options (But NOT Save Them).                    │
│ [F3]  - Use New Options and Save Them as Defaults.              │
│ [Esc] - Cancel.                                                 │
└─────────────────────────────────────────────────────────────────┘
```

Figure 41: *The Scale Options dialogue box in Design CAD.*

Set Resolution of Measurement

The **SO** command (or **Options | Scale Options**) lets you define precision and angular measurement for decimal, fractional, and engineering formats:

- **Decimal:** precision ranges from 1,000,000 to 0.00 000 01
- **Fractional:** precision ranges from 1,000,000 to 1/128

Set Style of Angular Measurement

The angular measurement options of the **SO** command are:
- 0 degrees is east and counterclockwise (mathematical), or north and clockwise (geographical)
- Degrees are displayed as decimal degrees (D.ddd), grads, radian, or degrees-minutes-seconds, DMS.ddd

Drawing Origin and Extents

You move the origin (0,0) of the drawing with the **Cursor | Origin** command. There is no command in DesignCAD to set limits or extents.

Set Object Snaps and Drawing Modes

DesignCAD does not have object snap modes in the manner of most other CAD packages. Instead, it has specific entity commands that snap to geometric features. Some examples are, as follows:

- **CT2** draws a circle tangent to two lines
- **CT3** draws a circle tangent to three lines
- **INT** snaps to the intersection of two lines
- **LS** snaps to the nearest line endpoint
- **PERP** draws a perpendicular to an existing line
- **TAN1** draws a line tangent to a circle

Similarly, there is no ortho mode. Rather, use the **OL** command to draw a vertical or horizontal line. After drawing the orthogonal lines, you rotate them with the **SRO** command (short for select rotate).

The **GO** command (or **Options | Grid Options**) lets you set the following options:

- Toggle snap-to-grid
- Set the snap increment
- Toggle display of the grid
- Set the grid increment
- Select style of grid: dots, axis marks, or ruler lines.

As an alternative, the **G** command toggles the grid display, while [Alt]-G toggles snap to grid. You cannot set the x- and y-increments independently. DesignCAD does not have an isometric mode.

Load Text Fonts and Define Text Styles

DesignCAD limits you to using a single font in a drawing; the work-around is to place text with the **Text Vector** command, which converts the text into vector entities (see Figure 42).

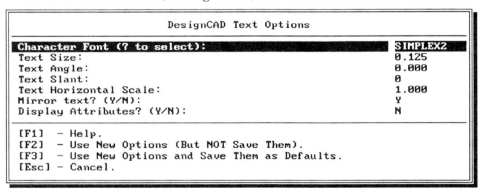

Figure 42: *The Text Options dialogue box in Design CAD.*

The **Options | Text Options** command sets the following parameters:

- Name of text font
- Height of text
- Angle of text line
- Width ratio
- Toggle mirrored text

The command also lets you toggle the visibility of attribute text. Place text with the **Text | Text** menu pick.

DesignCAD comes with 18 fonts, each in its own VFN font definition file. You can create custom fonts.

Load Linetypes and Set Linetype Scale

DesignCAD has eight hardwired linetypes; you cannot create custom linetypes (see Figure 43).

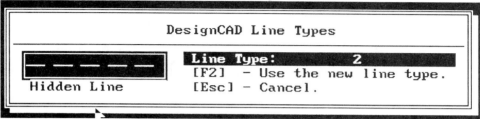

Figure 43: *The Line Types dialogue box in Design CAD.*

You set the working linetype with the **Options | Line Type** menu pick. As an alternative, the **Options | Drawing Opts** command lets you select a new linetype and specify the scale factor (see Figure 44).

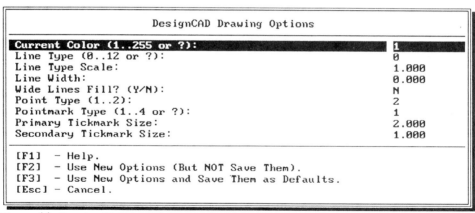

Figure 44: *Drawing Options dialogue box in Design CAD.*

Create Layers

DesignCAD supports 256 layers, numbered from 0 to 255; optionally, you can assign each layer an eight-character name (see Figure 45).

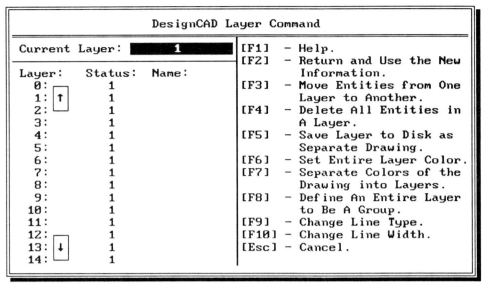

Figure 45: *The Layer Command dialogue box in Design CAD.*

The **Options | Layer | Layers** command lets you perform the following functions:

- Set the current layer
- Toggle a layer between active and inactive
- Name a layer
- Set the color of a layer
- Set the linetype of a layer
- Set the line width of a layer
- Move objects from one layer to another
- Save contents of a layer to a drawing file on disk
- Define contents of a layer as a group
- Delete all objects on a layer.

A powerful feature is this command's ability to separate objects of common color into layers. This is particularly useful after reading in an HPGL plot file. The **LD** and **LE** commands disable and enable layers.

Load Hatch Patterns and Pattern Scale

DesignCAD comes with 47 hatch patterns defined in a single file, DcHatch.Sys. You can create custom hatch patterns.

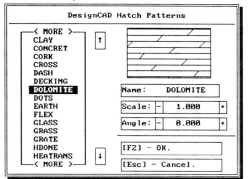

Figure 46: *The Hatch Patterns dialogue box in Design CAD.*

The **Draw | More | Hatch** command displays a dialogue box with the following information (see Figure 46):

- Select hatch pattern name
- Set the scale
- Set the angle of hatching

You specify a new origin for the hatching with the **Cursor | Origin** command.

Load Symbol Libraries and External References

To create symbols in DesignCAD takes two steps:

1. Create a group of objects with the **Groups | Groups Define** command.

2. Save the group to a DW2 file on disk with the **Select Save** command.

You load and place the symbol with the **Files | Symbol Load** command.

DesignCAD supports a form of externally reference files, called an "overlay." An overlayed drawing is only viewed and cannot be manipulated. Overlay a drawing with the **Files | More | Overlay** command; remove the overlay with the **Regenerate** command.

Set Dimension Variables and Scale

Dimension styles are set via two commands in DesignCAD: **Options | Dimension Opts | General Opts** and **Special Opts** (see Figure 47). The general options include:

- Toggle tolerances
- Select arrowhead
- Toggle associative dimensions (called "dynamic")
- Set dimension color
- Specify layer for dimensions
- Style of extension line, overshoot, gap, and length.
- Text size and scale

```
             DesignCAD General Dimension Options
 Enable Tolerancing? (Y/N):                              N
 Terminator (Arrowhead) Type (1..12 or ?):               1
 Enable Dynamic Dimensioning? (Y/N):                     Y
 Default Dimension Color (0..255, 0=none, or ?):         0
 Default Dimension Layer (0..255, 0=none):               0
 Extension Line Format (1=Variable, 2=Fixed Length):     1
 Dimension Text Size (0=Same as Regular Text):           0.000
 Dimension Text Horiz. Scale (0=Same as Regular Text):   0.000
 Dimension Extension Line Overshoot Length:              50.000
 Dimension Extension Line Gap size:                      100.000
 Dimension Extension Line Length:                        200.000

 [F1]  - Help.
 [F2]  - Use New Options (But NOT Save Them).
 [F3]  - Use New Options and Save Them as Defaults.
 [Esc] - Cancel.
```

Figure 47: *The General Dimension Options dialogue box in Design CAD.*

The special options are:
- Orientation of text for progressive dimensions
- Distance between text in progressive dimensions
- Size of balloons

Both menus save the options in the drawing; however, you cannot save and load dimension styles.

Select Default Plotter and Set Plot Style

The **Files | Plot** command sets plot parameters and plots the drawing (see Figure 48). A special feature of DesignCAD lets you plot a drawing over multiple sheets of media.

```
                    DesignCAD Print Command
     Density (1=Normal, 2=High, 3=Very High):      3
     Requested Drawing Units per Inch:             0.000
     Printing Area Width, inches:                  2.425
     Printing Area Length, inches:                 10.000
     Paper Width (?-help), inches:                 8.500
     Paper Length (?-help), inches:                11.000
     Rotate Drawing 90 Degrees? (Y/N):             N
     Center Drawing? (Y/N):                        Y
     Panel Mark (0-None, 1-Corner, 2-Box):         0
     Number Panels? (Y/N):                         N
     Character Font (? to select):                 SIMPLEX2
     Number of Copies to be Printed:               1
     Print the Attributes? (Y/N):                  N
     Print to Disk? (Y/N):                         N

            Use ARROW keys to select an item to be changed.
   F1-Help, F2-Print, F3-Print and Save Current Options, Esc-Cancel.
                Drawing size: 0.126 X 0.520 drawing units.
```

Figure 48: *The Print Command menu screen in Design CAD.*

Save Prototype Drawing

While DesignCAD does not support prototype drawings, you can save a drawing with the default settings, load and rename it with the **Files | Retrieve** and **Files | Save** commands. The process can be automated by defining a macro called AutoExec.

The **Options | Config Save** command saves many drawing parameters to the DCad2.Sys setup file on disk.

AutoSketch for Windows

The concept of multiple prototype drawings does not exist in AutoSketch. The work-around is to create and save a pseudo-prototype drawing. Load the prototype drawing then immediately save it by the working name.

Open a New Drawing

Start AutoSketch in Windows by double-clicking on the AutoSketch icon. When you start AutoSketch without a drawing, it automatically creates a new, blank drawing named Unnamed.Skd (see Figure 49).

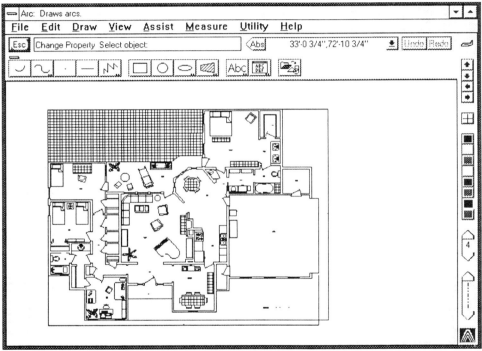

Figure 49: *AutoSketch for Windows runs under the Windows operating environment.*

Set Unit of Measurement

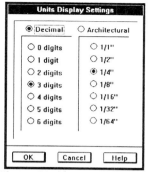

Figure 50: *The Units Display Setting dialogue box in AutoSketch.*

AutoSketch uses an unitless mode of recording drawing measurement. The **Utility | Drawing Settings | Units** command combines setting the unit and resolution of measurement (see Figure 50):

- **Decimal:** 0 to 6 decimal places of precision
- **Architectural:** 1" to 1/64" precision

The decimal units are suitable for metric use.

Set Style of Angular Measurement

AutoSketch does not let you select a style for displaying angles. All angles are reported in D.ddd (decimal) format; the number of decimal places are controlled by the **Utility | Drawing Settings | Units** command.

Drawing Origin and Extents

While you cannot change the origin of the drawing, you can change the origin of symbols (called "parts" by AutoSketch) with the **Utility | Drawing Settings | Part Base** command.

The drawing extents are set with the **Utility | Drawing Settings | Limits** command. This affects the extent of the grid display and the Zoom Limits command.

Set Object Snaps and Drawing Modes

AutoSketch has eight object snap modes found in the **Assist** menu:

- Node
- Endpoint
- Midpoint
- Perpendicular
- Intersection
- Center
- Quadrant
- Tangent

When you toggle on any modes, they remain on until you turn them off. The **Assist | Attach** command (or press [Alt]+F8) turns all selected modes on and off as a group. In addition, the **Assist | Dim Slide** command forces dimension text to move only parallel to the dimension line.

Set the size of the grid and snap increments with the **Utility | Drawing Settings** menu. Toggle the grid, snap, and ortho modes via the **Assist** menu. You can specify independent x- and y-sizes for the grid and snap increment with **Utility | Drawing Settings | Grid/Snap/Limits** command; grid and snap settings are independent. You cannot rotate the orthogonal angle. AutoSketch does not have an isometric drawing mode.

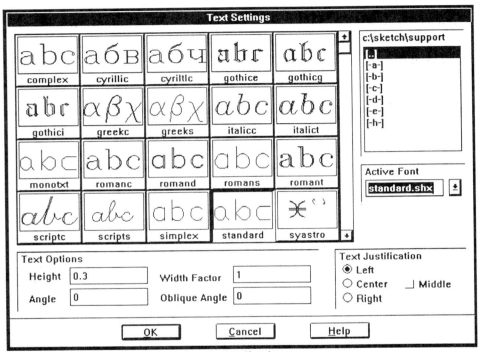

Figure 51: *The Text Settings dialogue box in AutoSketch.*

Load Text Fonts and Define Text Styles

The **Draw | Text Editor | Settings** command combines selecting the font and text style. The Text Settings dialogue box lets you set the following parameters (see Figure 51):

- Load and select font
- Text height
- Angle of text line
- Character width
- Character angle
- Text justification

You can have only one text style active in an AutoSketch drawing.

AutoSketch comes with 24 text fonts stored on disk in individual SHX files. You can create custom text fonts and read AutoCAD SHP text font files.

Load Linetypes and Set Linetype Scale

AutoSketch has ten hardwired linetypes; you cannot create custom linetypes. The **Utility | Drawing Settings | Line Type** command sets the working linetype and scale factor (see Figure 52).

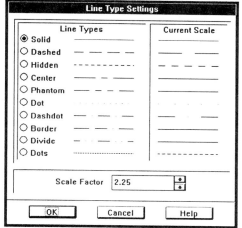

Figure 52 *The Line Type Settings dialogue box in AutoSketch.*

Create Layers

AutoSketch has ten layers numbered 1 through 10; you cannot name layers. The **Utility | Drawing Settings | Layer** command lets you set the working layer number and toggle the visibility of layers (see Figure 53). Invisible layers cannot be edited or plotted.

You cannot set attributes for a layer, such as color or linetype. Instead you set color (of 255 available) independently with **Utility | Drawing Settings | Color**.

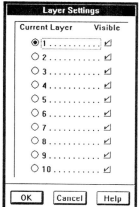

Figure 53: *Layer Settings dialogue box in AutoSketch.*

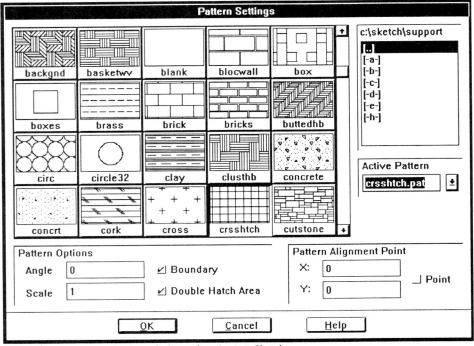

Figure 54: *The Pattern Settings dialogue box in AutoSketch.*

Load Hatch Patterns and Pattern Scale

AutoSketch comes with 55 hatch patterns saved in individual PAT files. You can create custom hatch patterns. The **Utility | Drawing Settings | Pattern** command selects the active pattern and the following parameters (see Figure 54):

- Angle
- Scale
- Pattern origin coordinates
- Toggle boundary
- Single or double hatch area

The **Draw | Pattern Fill** command loads the PAT file and fills circles and closed polygons the pattern selected as you draw them.

Load Symbol Libraries and External References

AutoSketch calls symbols "parts." You save a portion of the drawing as an SKD part file on disk with the **File | Part Clip** command. Place the part with the **Draw | Part** command. The AutoSketch package includes 2,000 parts. AutoSketch does not reference external drawings.

Set Dimension Variables and Scale.

AutoSketch supports a single dimension style in the drawing, set with the **Utility | Drawing Settings | Dimension** command. It lets you control the following parameters (see Figure 55):

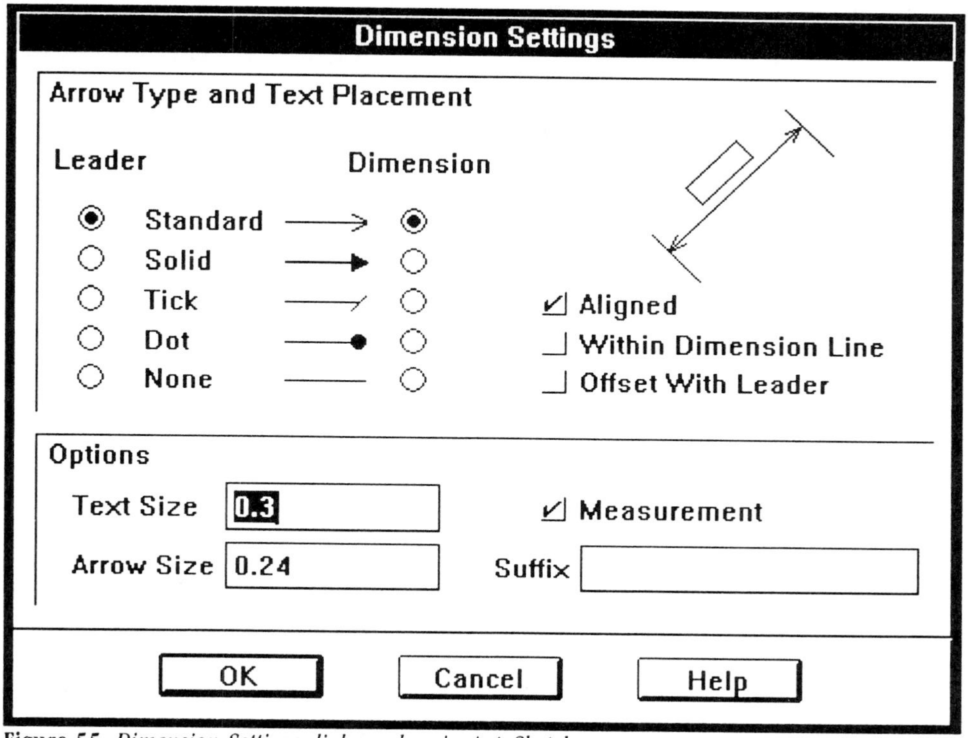

Figure 55: *Dimension Settings dialogue box in AutoSketch.*

- Arrow type
- Arrow size
- Text alignment
- Text size
- Toggle associativity
- Suffix text up to 19 characters

To change the dimension style after placement, use the **Edit | Property** command.

Select Default Plotter and Set Plot Style

AutoSketch uses the Windows system printers for output. Use the **File | Print Settings** command to set the plot parameters; these remain in effect until changed again. The **File | Print** command prints the drawing.

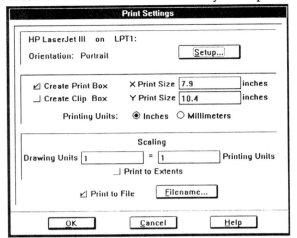

Figure 56: *The Print Settings dialogue box in AutoSketch.*

Save Prototype Drawing

While AutoSketch does not support prototype drawings, you can use a start-up macro to automatically load a template drawing; then use immediately rename the drawing with the **File | Save As** command.

Many of AutoSketch's settings are found in the Sketch.Ini initialization file. Edit the file with a text editor, then save it in different subdirectories. Drawing settings can be changed with the Set command within AutoSketch, including the drawing directory with the DrawingDir variable.

Summary

This chapter gave you hands-on experience for creating a prototype drawing for eight CAD packages. The next chapter describes how to create a printed standards manual.

CHAPTER 9

Writing a CAD Standard

With the standards in place, you should document them on paper in a three-ring binder. In this form, the CAD operators can quickly scan through the standards; the binder can be given to clients and contractors. This chapter illustrates a typical table of contents approach to writing the office CAD standard.

The Office Standards Manual

The office CAD standards manual is a three-ring binder that contains the following sections:

- Title page
- Disclaimer page
- Table of Contents
- Standards
- Index

Title Page

The title page identifies the documentation as the office CAD standards manual. The title page should include the following information (see Figure 1):

- Company or agency name and address
- Title, such as "Standard for Electronic Documents," "CAD Procedures Manual," "Consultant Guidelines for CAD Drawings," or

> **Executive Summary**
>
> The office CAD standards manual consists of the following parts:
>
> - Title page
> - Disclaimer page
> - Table of Contents
> - Standards
> - Index
>
> The standards of the manual should include the following sections:
>
> - Purpose
> - Creating a Drawing
> - Plotting the Drawing
> - Drawing Management
> - CAD Procedures
>
> To help keep the documentation up to date, use a full-featured word processor's automation tools. ■

"CAD User Manual of Instruction."
- Revision number
- Most-recent revision date

Every page of the documentation should have the revision number as a header or footer.

Disclaimer page

The disclaimer page follows the title page and announces to readers and users that your firm is not responsible for problems arising from use of the standard. You may also want to prohibit the sale and modification of the standards.

Since CAD standards are still emerging, it pays to share your standards with other firms. That helps improve the standardization of the CAD industry and its drawings.

Neenah, Delbert & Associates
123 Main Street
Yourtown, USA 98295-3053
(604) 555-1212

CAD Procedures Manual

Consultant Guidelines for CAD Drawings

Revision 4
Created: 10 October 1990
Last revised: 24 September 1993

Figure 1: *The title page of an office CAD standards manual.*

Table of Contents

The table of contents forms the outline of the entire documentation. Here is a outline you can use as a guide for your standards book:

1. **Purpose**
   a. Where to Go for Help
   b. Sources of Documentation
   c. Monthly User Group Meeting

2. **Creating a Drawing**
   a. Filenames
      i. *Drawing Log*
      ii. *Drawing Management*
   b. Sheet Numbers
      i. *Title Blocks*
      ii. *Sheet Borders*
   c. Units
      i. *Dimensions*
      ii. *Elevations and Datum*
   d. Coordinates
      i. *Grid Lines*
      ii. *Match Lines*
      iii. *Reference Marks*
   e. Use of Symbols
      i. *Reference Tags*
      ii. *Standard Office Library*
   f. Symbol Naming
   g. External References
   h. Layers
      i. *Names*
      ii. *Colors*
      iii. *Linetypes*
      4. Standard Office Library
   i. Hatch Patterns
      i. *Scale Factors*
      ii. *Standard Office Library*
   j. Line Widths
   k. Text
      i. *Fonts*
      ii. *Notes*

3. **Plotting the Drawing**
    a. Plot Scales
    b. Sheet Sizes
    c. Pen Plotter Use
    d. Laser Printer Use
    e. Batch Plotting
    f. Network Plotting

4. **Drawing Management**
    a. Backups
    b. Archives
    c. Accessing the Network
        i. *Network Schematic*
    d. The Menu Systems
        i. *Screen Menu*
        ii. *Icon Menu*
        iii. *Tablet Menu*
    e. File Extensions
    f. Translating Drawings
        i. *Via DXF*
        ii. *Via IGES*
        iii. *Via DWG*

5. **CAD Procedures**
    a. Flow Diagram for Project Drawings
    b. Adding to the Symbol Library
    c. Adding Hatch Patterns and Linetypes
    d. Function Key Definitions

6. **Appendices**

It is better to be terse and symbolic, rather than long winded in the text. Remember that this is a reference that will be scanned by employees; no one will sit down and read it from cover to cover.

Standard Office Library

Not only should the office standards be embedded in a prototype drawing, but the standards should be liberally illustrated throughout the documentation.

Figure 2 illustrates how to set up a page that documents visual elements. This can be applied to linetypes, the symbol library, hatch patterns, and text fonts, if more than one font is used.

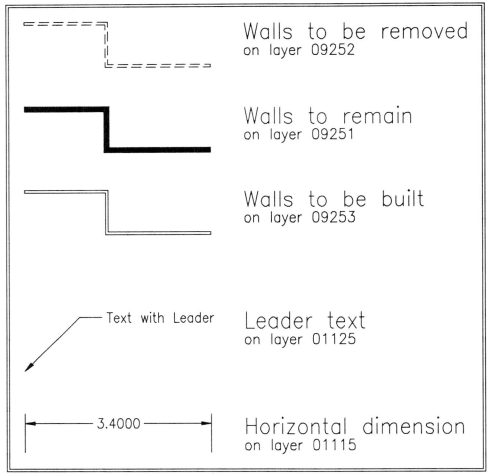

Figure 2: *Documenting visual elements of a CAD drawing.*

After the Document is Complete

After completing the CAD standard documentation, take a break!

But you will soon find that the document is a living standard. From time to time, you will have to update it. I recommend writing the standard with a high-end word processor, such as WordPerfect and Microsoft Word. Use the word processor's automation to keep the document up to date, as follows:

- Use the spell checker and grammar checker to produce a well-written document. Spelling mistakes and bad grammar reduce the credibility of your office standards.

- Use the Table of Contents and Index generators to automatically renew the table of contents and index.

- If you are working under Windows (or OS/2), use OLE (or SOM) to create links between the graphics and the document.

- For low-cost distribution of the standard, maintain an electronic copy on diskettes.

Summary

This chapter described how to write down on paper your office CAD standard in an organized manner that is readily accessible to the users. In the next section of this book, we look at how to maximize the efficiency of CAD in your office.

SECTION III

Maximizing CAD Efficiency

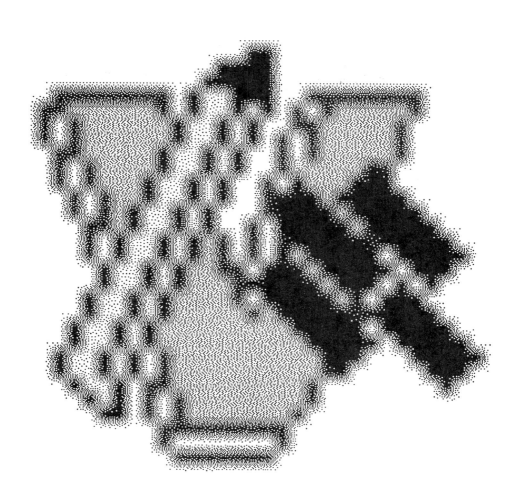

CHAPTER 10

Input, Output

Many of the previous chapters (indeed, most books on CAD) tell you how to create drawings in CAD. What about the drawings that already exist on paper? And what's the best way to get CAD drawings back onto paper? This chapter describes strategies for dealing with paper and provides a rationale for *not* dealing with paper.

The Electronic Drawing

It is one of my favorite applications stories and the first time I heard about the all-electronic office where no drawings exist on paper—from initial concept to finished product. I enjoy telling this story to new clients to help them understand the potential of CAD; usually, their mouths drop open as I tell it. Here's the story:

A log home builder in my neck of the woods implemented AutoCAD almost a decade ago. At first, it was merely a replacement for the drafting board. But the firm has been lucky enough to hire a CAD operator who had grand ideas for what CAD could do. Through his efforts, the log home builder eliminated paper drawing from the design construction process. To see how this is possible, let's follow the process from conceptual design through to finished product.

A customer is interested in having a log home built on a remote island site. When she comes into the builder's office, the salesman invites her to sit down at a computer with a CAD operator. The operator brings up renderings of past log home projects. When the customer sees a home that catches her fancy, the CAD operator brings up the floor plan. As the customer discusses the changes she would like with the salesman, the

> # Executive Summary
>
> *Most firms produced drawings on paper before CAD. The approach to dealing with paper drawings depends on how your firm makes use of them:*
>
> - Ignore the existing stock of paper drawings
> - Archive the drawings in digital format
> - Convert the drawings wholesale to vector format
> - Use the drawings as a raster background to vector-based revisions
>
> In the three latter cases, you need a D- or E-size raster scanner to read the drawings into a raster file format understood by computer software.
>
> Consider eliminating paper drawings from your firm. If your clients or other departments in your firm work with digital data, chances are you will find a way to eliminate the need for paper drawings. ■

CAD operator enters them into the electronic drawing.

With the final changes in place, the CAD package generates a 3D view from the 2D floor plans, then renders the wireframe drawing to let the customer appreciate the proposed design changes. With the customers approval, the CAD software generates three lists:

- A price list with two prices (manufacturing cost and retail price) is printed in the showroom. The customer and salesman haggle over the final selling price.
- A log cutting list is transmitted to the log yard. In the log yard, the cutting list (in NC numerical control code) instructs the saws to cut the wood needed for the house.
- A parts list of non-log items (windows, door knobs, et cetera) is transmitted to the warehouse.

The CAD system also generates assembly drawings (on paper) for the site. When all the parts are together, they are airlifted by helicopter to the island site where the home is assembled.

Eliminating the Paper Trail

The log-home story was just the first time I'd heard of all-electronic drawings; since then, I've heard many more. The most common application is with lathes and other metal milling machines, which are typically computerized. Third-party software exists that converts a CAD drawing into the NC code that instructs the machines.

Another example is survey data, which at one time was written laboriously in waterproof notebooks. Survey instruments now record data electronically in their own memory banks. At the end of the day, the data is dumped to the on-site computer or transmitted via modem back to the office. Third-party software translates the survey data into a CAD drawing. The exciting part of all this is that *no paper drawings are used* in the manufacturing process, which eliminates:

- The time needed to plot drawings on paper
- The errors made in reading data
- The expense of re-entering data from paper sources (notebooks, other drawings) into the computer

When you set up the CAD system, consider whether paper drawings can be eliminated entirely, all except for check plots. If your clients or other departments in your firm use data in digital form, then chances are you will find a way to eliminate the need for paper drawings. Here are some ways to think about working with all-electronic drawings:

1. Hand clients your drawings on diskette or tape.

2. Establish an account with an on-line service (such as Internet or CompuServe) to communicate messages and data via electronic mail between your office, home-based employees, and your clients.

3. Send data back from the field via modem and remote-control software. Notebook computer modems can now use cellular phones.

4. Install a dedicated serial connection between the office computers and shop floor machinery. Often a 9600 baud line is all it takes.

5. Set up a LAN (local-area network) to share drawings and other electronic data within your office.

6. Set up a WAN (wide-area network) to share electronic data throughout the entire organization, no matter where in the world they are located.

If you've decided that your office still needs to deal in paper, then read on. After all, as John Dvorak once said:

> *"The paperless office is
> as likely as a paperless bathroom."*

The Input Options

Most books on CAD concern themselves with tutoring the reader in creating a drawing with computer software. However, the majority of drawings in existence are on paper. For example, one of the world's navies has reportedly has entire warehouses filled with paper drawings. Why do you need to worry about non-CAD paper drawings? Here are two reasons:

- The media is slowly disintegrating, particularly drawings drafted on acidic paper. Drawings created in the 1950s or earlier may have crumbled into flakes of paper by now.
- You may need to reuse existing site and facility drawings for new projects and renovations.

Knowing the two reasons for worrying about paper-based drawings helps lead you to the appropriate solution: (1) archiving, and (2) conversion.

Archiving Drawings

If your worry is about flaking paper, then you are concerned about maintaining the drawings as records. To preserve your firm's investment in drawings, the solution is to scan the drawings electronically and store the data on opto-magnetic media. Archiving is trivial but dull. One hydro power authority estimated it would take until the year 2020 for a single person to scan all of their drawings into electronic format.

The technology for this process is readily available and reasonably priced. Here's what you will need:

1. **An E-size scanner** that reads the drawings. Today's scanners read an E-size drawing at 200 dot per inch less than a minute.

2. **Scanner software** that reads the output from the scanner and adjusts the scan for minor imperfections, such as skew and detritus. The scanner usually outputs raster data in RLE (run length encoding) compressed format. The software saves the scanned image in a file on disk, usually in a variety of raster formats such as TIFF and PCX.

3. **Archival software** that archives the image files in compressed format on the backup device.

4. **A backup device** that stores the compressed image files on the archival media. The choice of media (regular tape cartridge, digital tape, or opto-magnetic disks) depends on your budget and need for capacity.

5. **Retrieval software** that lets you select, view, and print archived drawings.

6. **A raster output device,** such as A-size laser printer or E-size inkjet plotter, to make a hardcopy of the archived drawing.

If you need to regularly access the archived drawings, your firm may need to consider a jukebox retrieval system where a robotics device finds and selects the tape or disc and inserts it into the reader. At the top end, IBM's Dataserver is a data archive system that uses 64 tape drives and a robotic arm to access nearly 20,000 tape cartridges holding 45TB (terrabytes) or 45,000,000 megabytes of data.

The advantages to digital archiving are: smaller storage needs than paper drawings; the plastic media is more stable than paper; and access to drawings is computerized and hence faster.

The disadvantages are: the drawings, while electronic, cannot be used in CAD; the true media lifetime of plastic and mylar is not known; the raster format is bulky and, hence, slow to load into software (an uncompressed E-size monochrome drawing occupies 16MB of disk space).

Converting Drawings

In the mid-eighties, programmers and their marketeers were sure that computerdom's most perplexing problem was close to being solved: feature recognition. Their trumpeting mislead many users—even to this day—that it is possible for software to convert a scanned drawing image into a perfect vector CAD file. Let's look at the problem more closely to understand why it is so difficult to solve.

When you look at a drawing, your eye and brain make out the lines, circles, and text as independent yet connected objects. You are able to read text stained with coffee; you understand lines with arrows attached to be dimensions; you know the difference between the lines making up the drawing's border and the line representing the building walls. Late in the twentieth century, computers still cannot recognize these features.

The reason computers do a poor job of feature recognition is that currently the only technology that "reads" a paper drawing is the raster scanner. Some work is going into developing a vector scanner, which at this point is limited to tracing contour maps.

The raster scanner contains a row of light sensors—some 200 sensors per inch of scanner width or 7,000 sensors for an E-size scanner. Some scanners use one or two CCD (charge-coupled device) lenses—the same lens used in video cameras—that scan the entire width of the drawing.

As the drawing passes through the scanner, it takes a reading from all sensors every split second—200 readings per inch of drawing passing through, from which comes the 200dpi resolution rating. With each reading, each sensor reports whether it detects light (the paper's background) or dark (a drawn line). Most scanners can also sense up to 64 relative levels of light and dark, called "levels of grey."

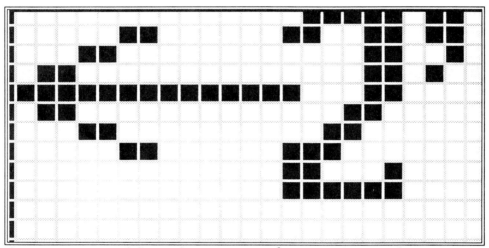

Figure 1: *A scanner reads lines and text as an array of 200 square dots per inch.*

The result of the data collection is a matrix of values, dark and light. Unfortunately, reading the drawing in 1/200" increments loses the vector information. Figure 1 shows how part of a 2" dimension is represented. You probably can see the extension line, the arrowhead, dimension line, the number "2" and part of the inch (") symbol because the brain excels at pattern recognition.

Now it is up to the computer software, called "raster to vector converters" to make sense of the dots. The software that exists today can determine lines and arcs. Some software recognizes text after a fashion by isolating all objects smaller than 40 pixels square. Very high end raster-to-vector conversion software claims to be able to reconcile dimensions.

Finally, the conversion of paper drawings into electronic format simply

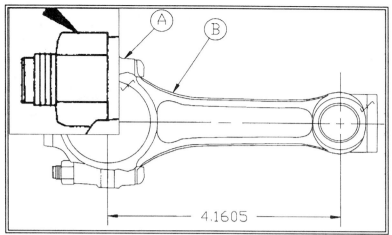

Figure 2: *A scan of a drawing looks jagged when enlarged.*

cannot recreate layers, attributes, colors, database links and even has difficulty with linetypes. In short, all the data that you expect to find in an electronic drawing cannot be recreated—nor will they ever be—through the raster-to vector conversion process.

In the Figure 2, the drawing of a crankshaft has been scanned in. In the upper left corner, the detail shows the nut enlarged four times. Note how seemingly straight lines are recorded as jagged and broken segments.

The scanned image was automatically converted into a vector format called DXB and read into AutoCAD with the **DxbIn** command. Figure 3, on the next page, shows the resulting detail of the nut; notice how the arrowhead is not fully formed and the uneven lines making up the bolt and screw thread. The full drawing shown in the upper right.

Original Drawings Are Not Accurate

We have seen that scanning and automatic conversion is not necessarily accurate. That leads to the other problem of inputing paper drawings into CAD: the inaccuracy of the original drawings. We have come to expect that drawings created with CAD are accurate to six or more decimal places. We tend to forget that drawings made with a pencil are inherently inaccurate. The accuracy of a handmade drawing is approximately the width of the pencil lead—or pen tip—about one decimal place of accuracy.

If you've been a manual drafter, then you know that a certain amount of fudging goes on: lines look straight but probably run at a slight angle to match up; corners that look square but probably are off by a degree or two; circles that appear to have an accurate radius but are probably undersized; dimensions might not measure up to their stated distance; and

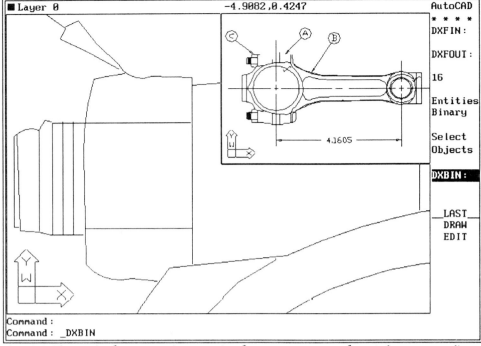

Figure 3: *The result after automatic conversion from raster to vector format shows uneven lines.*

fillets tend to not quite join at the corner.

On top of the inherent inaccuracy of manual drafting, it was common in pre-CAD days for the drafter to declare in the scale portion of the drawing's title box: "Not to Scale." The drawing looked good but the drafter had made no attempt to accurately scale the parts.

Thus, there is no point in having really accurate raster-to-vector conversion software if the original paper drawings are inaccurate.

The Two Editing Solutions

We have seen the difficulty of inputing paper drawings into a CAD system. The problems are inherent in every step of automatic conversion:

- The original paper drawing is inaccurate due to manual drafting.
- The scanner reads paper drawings as an array of meaningless dots.
- The raster-to-vector conversion software finds rudimentary patterns among the dots.
- The vectorized drawing requires extensive post-conversion editing within the CAD system.

With these drawbacks to automatic conversion, what are the alternatives? Here are four ideas:

1. Use automatic conversion if the accuracy and intelligence of the drawing are not an issue in your discipline.

2. Use automatic conversion for small jobs after scanning client logos and parts from vendor catalogues.

3. Use text recognition software to convert raster text into ASCII format for input into word processing software.

4. Use the primary alternative to automatic conversion—manual digitizing. The paper drawing is taped to an E-size digitizing board and a drafter traces over the drawing, applying human intelligence to feature recognition.

Some firms employ speech input to help speed the drafting process. This lets the drafter speak commands, rather than type them in at the keyboard.

The Partial Conversion Strategy

When it was recognized in the late 1980s that software was incapable of converting raster scans into vector drawings, Robert Godgart came up with a radical re-engineering of the process: if raster images cannot be accurately converted into vector format, then don't convert them.

He rationalized that most firms don't actually want to convert an entire paper drawing into vector. Instead, most firms use the old drawings for renovation work—perhaps changing the size of a room or rerouting the path of a highway. His re-engineering went as follows:

1. Keep the raster image in raster format.
2. Display the raster drawing as a background image within the CAD system.
3. Erase the portions of the raster drawing that no longer apply.
4. Draw the new and modified parts with the CAD package's vector commands.
5. Convert the vector lines into raster format (a trivial exercise).
6. Print out the revised drawing on a raster output device, such as laser printer or inkjet plotter.
7. Store the revised drawing in raster format again.

The approach to take to dealing with paper drawings depends on how your firm uses them. You can either ignore them, archive them in digital format, attempt to convert them wholesale to vector format, or use them as a raster background to vector-based revisions.

The Output Options

Your firm may be able to ignore the problem of dealing with existing paper drawings but outputting CAD drawings on paper is a challenge faced by every CAD user—unless, of course, your firm has gone all-electronic, as described at the beginning of the chapter.

Plotting drawings is one of the drawbacks to CAD. There are so many parameters involved in plotting that it isn't a surprise that the plot comes out looking wrong.

Some problems can be solved before the plot takes place. Use the CAD software's plot preview feature to see how the drawing will be placed on the media. The plot previewer draws a rectangle representing the paper's margins around the drawing. You then move and size the rectangle to fit and scale the drawing correctly.

Here are plotting problems that occur (along with the possible cure):

Problem: *In manual drafting, when the drafter is finished with the drawing, the drawing is finished and is immediately ready for blueprinting or archiving. In CAD drafting, when the drafter is finished, the drawing still needs to be plotted on paper.*

Solution: The CAD manager must allow additional time at the end of the project for plotting drawings. The time to plot a CAD drawing ranges from over one hour for a pen plotter to under five minutes for an electrostatic plotter. Thus, large projects can take hundreds of hours of plotting time.

Note that raster plotters (electrostatic, thermal, inkjet, dot-matrix, and laser) have the ability to produce additional copies of the plot in speeds ranging from a couple minutes to 17 per minute—thus obviating the need for blueprinting. If meeting a deadline becomes a problem, contract a service bureau to produce plots for you (see Chapter 11).

Problem: *While the manual drafter works on the drawing, other staff and managers can easily view the progress sine the entire drawing is on display. In CAD drafting, the drafter is usually zoomed in on a detail, obscuring the bulk of the drawing.*

Solution: The work-in-progress can be viewed in two ways: electronically and mechanically. On a network, the CAD manager can view the progress of drawing with viewing software. This software can display up to 25 drawings at a time from any subdirectory on the network, and lets the manager markup the drawing via redlining features.

As an alternative, the CAD operator regularly (such as at the end of each shift) creates a check plot on an A- or B-size laser printer.

Problem: *In manual drafting, it's easy to make a change after the drawing is finished: just get out the electric eraser, ink in the revision, and a revision note. With CAD, you always have two originals of a drawing: the electronic original on disk and the plotted original on paper. If you change the electronic drawing, you must replot the entire paper drawing; if you change the paper drawing, you have to go back into the electronic drawing and duplicate the changes.*

Solution: Pencil plotters are available now that let you easily make manual changes to plotted drawings. Solvent-impregnated erasers help with ink plots.

Put in place a workflow system that ensures the electronic version of the drawing always matches the plotted copy. Some firms keep the drawings on diskette; whoever holds the diskette has ownership of the drawing. Most firms have a drawing log sheet, managed either electronically as a front-end to the CAD package or on a paper form.

Problem: *Networking a CAD system limits the number of stations available for plotting. If all the network licenses are used up, none may be available for plotting.*

Solution: Most CAD systems have a separate plotting utility: MicroStation, CadKey, and DesignCAD. AutoCAD has a "free plot" mode that lets you run one or more copies just for plotting without using up a network license.

Windows-based CAD packages can use the Windows Print Manager to plot as a background task. The Print Manager captures the output from the CAD software and keeps it in a file on disk. This process occurs fairly quickly. Then the CAD system is free to do other work, while the Print Manager feeds the disk file to the plotter or printer. Third-party vendors make plot spooler software available that performs the same function under DOS.

Problem: *Plotting ties up workstations.*

Solution: Most operating systems—even DOS—are able to plot "in the background." That means a small program (called a "spooler") sends the plot data to the plotter while you continue to edit and draft with the CAD software. This multi-tasking works by the spooler stealing small, almost imperceptible amounts of time from the CPU.

If using a spooler is not possible in your system, then consider dedicating a single computer to tend the plotter, known as a "plot server." This computer can be the least powerful one in the office, since sending plot data to a plotter is a task even the original 8088-based PC/XT is capable of.

Problem: *Nothing happens.*

Solution: The most common causes are:

- Cables are not tightly fastened; use the screws on cable connectors.
- CAD software is using the wrong plotter driver; check the plotter and CAD documentation for plotter driver compatibility at both end, such as HPGL, HPGL/2, DM/PL, and PCI.
- Incorrect communications parameters are set in the CAD software or in the plotter's hardware, such as the wrong baud rate or communications port.
- The plotter is off-line; press the plotter's [On Line] button and ensure the On Line light turns on.

Problem: *Edges of the plot are missing.*

Solution: All plotters—whether raster or pen—have a non-printable area along all four edges of the media where the plotter grips the paper. These non-plotting edges are called "margins." If you set the margins too small, the drawing is scaled too large and the plotter cannot draw. Check the plotter's documentation and reset the margins in the CAD software.

Problem: *Plot placed off-center on paper.*

Solution: Check the location of the plot origin in the CAD software. Or, you may be using the wrong plotter driver; for example, smaller Hewlett-Packard plotter have their plot origin in the center of the paper, while larger HP plotters have the origin at the lower-left corner.

Problem: *Plot created at wrong scale.*

Solution: Check the scale factor in the CAD software you specified for plotting. Or, you may have specified the wrong media size. If nothing else, set the scale factor to "Fit," which makes the CAD software scale the drawing to the paper in the plotter.

Problem: *Plot drawn with wrong colors, line types, or widths.*

Solution: Check the pen matching. Make sure you have not specified hardware or plotter linetypes, where the plotter uses its linetypes.

Problem: *Lines are broken up.*

Solution: Pen may be skipping due to too-high plotting speed. Or, pen may be low on ink. Or, you may have specified hardware linetypes.

Problem: *Line looks blotchy or dried out.*

Solution: Wrong combination of ink and paper. Check the paper vendors recommendation for pen types.

Problem: *Text and dimensions are too small or too large.*

Solution: Change the scale factor in the drawing; do check plots during the drafting process to ensure text comes out the right size.

Problem: *Check plots take too long.*

Solution: Reduce the raster resolution: much of the drawing is legible at 150dpi on A-size paper yet the plot time is much faster than 300dpi.

Problem: *Parallel lines are not parallel.*

Solution: The plotter's accuracy is low. The plotter may need recalibration or repair or replacement.

Problem: *Objects appear on the plot that were not meant to be plotted.*

Solution: Layers were not turned off or frozen. Some versions of AutoCAD have a bug that plots non-plotting layers.

Problem: *Output looks ragged.*

Solution: You specified too low a resolution on a raster plotter. If the problem occurs on a pen plotter, the plotter's arc angle setting may be too coarse.

Problem: *Plotter spits out paper before starting plot.*

Solution: The wrong size of paper was specified in the plotter or CAD software.

Problem: *A second drawing is plotted on top of the first drawing.*

Solution: You haven't told the spooler to pause after each plot on a sheet-fed plotter.

Summary

If your firm must deal with paper, then this chapter described the problems and solutions to inputing and outputing drawings on paper. In the next chapter, we look at working with a service bureau.

CHAPTER 11

Working With a Service Bureau

The service bureau can be considered a type of "office overload" service. When your plotters aren't keeping up with project deadlines or a client dumps fat rolls of paper drawings in your lap, get help from a service bureau. In addition, service bureaus provide one-of services when it is too expensive to purchase the equipment for your office, such as VCR recording of animations and loaner equipment.

What is a Service Bureau?

A service bureau provides CAD services when your own firm cannot. You may find your firm's plotters aren't going to produce hardcopy drawings in time for the deadline, or you've been given a job that requires an E-size digitizer, or the next release of CAD has been installed....

For extra help with your CAD work, every major city has dozens of service bureaus that help out in these "office overload" situations. Service bureaus are located in computer dealerships, in blueprinting houses, or are run by private consultants. Some bureaus specialize in a single aspect of CAD (digitizing or plotting), while others attempt to provide all possible services.

Before committing to a service bureau, check their references and look over their premises, More importantly, do a trial run *before* the crunch arrives. While most service bureaus do their best job for you, some are opened by recently-unemployed CAD drafters who lack the high-quality equipment needed to do a professional job.

Executive Summary

A service bureau can be considered a type of "office overload" service. Service bureaus provide one-of services when it is too expensive to purchase the equipment for your office. A typical service bureau offers some of the following services:

- Training in CAD software and computer hardware
- Customizing the CAD system for your discipline
- Programming add-on applications to your CAD package
- Scanning drawings electronically
- Digitizing drawings by hand
- Plotting drawings
- Translating drawings from one format to another
- Setting up a document management system
- Repairing specialty hardware
- Rendering and animation from 3D CAD files
- Color slide, video tape, and hardcopy output
- Loaner equipment
- Contract drafters

Rates vary by the hour (programming and customization), by the course, and by the square-foot (plotting and digitizing). You will find service bureaus at local computer dealers, blueprinting houses, and full-service, stand-alone bureaus. ■

If your location has no nearby service bureau, high-speed modems and overnight couriers make it possible to deal with an out-of-town service bureau. You can send the drawing file to the service bureau by modem and receive the plotted media back by Federal Express. Using data compression and a 9600 baud modem, it takes about seven minutes to transmit a 1MB drawing file over the phone lines (CAD drawings compress to about 40% of their original size). Some bureaus use off-shore labor—particularly for digitizing—to help keep their prices down.

Bureau Services

Here is a rundown of the services provided by CAD bureaus:

Input
- Scanning drawings electronically
- Digitizing drawings by hand
- Translating drawings from one format to another
- Contract drafters

Output
- Plotting drawings
- Rendering and animation from 3D CAD files
- Color slide, video tape, and hardcopy output

Maintenance
- Training in CAD software and computer hardware
- Customizing the CAD system for your discipline
- Programming add-on applications to your CAD package
- Setting up a document management system
- Repairing specialty hardware
- Loaner equipment

If the bureau offers training, check that they are recognized by the CAD company. For example, Autodesk has its ATCs, about 200 Authorized Training Centers in North America.

Drafters can become certified through the National Association of CAD/CAM Operators, with a certificate from an authorized training center or local college, or through the AutoCAD Certification Examination administered by Drake (see Appendix D).

Consultants can be certified in MicroStation through the Intergraph Registered Consultant program.

Service Bureau Pricing

The prices charged by service bureaus vary by bureau within cities and between cities due to differences in competition for service. You should expect to pay more in metropolitan areas on the coasts than in the middle of the continent.

Rates also differ depending on the service. The rates are lower for plotting (more automated) than for digitizing (more person-hours involved).

CAD Training Rates

Chapters 1 and 3 suggested approaches to training your staff. Here are sample prices you can expect to pay for CAD training:

- **One day training:** $300 to $400 per person
- **Three days training:** $700 to $900 per person

Generally, the lower price range is for basic CAD, while the higher prices are for advanced topics, such as programming, 3D Studio, and third-party applications.

Digitizing Drawings Rates

For digitizing, the price depends on the drawing size and the style of digitizing. The digitizing service costs more because it is more labor intensive and includes providing the drawing on diskette and with a check plot. Here is the range of prices charged by service bureaus:

- **Per hour:** $20 to $40 of digitizing
- **A-size:** $50 to $60 per digitized drawing; $5 for scanning only
- **D-size:** $100 to $250 per digitized drawing; $60 for scanning only
- **E-size:** $200 to $300 per digitized drawing; $85 for scanning only

Plotting Drawings Rates

For plotting, the price depends on the size of plot. Here is the range of prices charged by service bureaus:

- **Hourly:** $20 to $60 per plotter hour
- **A-size:** $6 to $15 per plot
- **D-size:** $20 to $45 per plot
- **E-size:** $30 to $70 per plot

The range of prices is due in part to the variance in pen and media prices. A technical pen on mylar is more expensive than a roller ball on plain paper.

In the case of plotting and digitizing, avoid paying by the hour since it is difficult to determine what the service bureau accomplishes in the hour. Instead, pay by the drawing.

Additional Charges

When looking for a service bureau, shop around for comparative prices. While I don't recommend sacrificing quality for price, you may find at least one bureau whose prices are far higher than the average.

When obtaining quotes, ask about additional charges. These might include:

- State and provincial taxes
- Federal taxes, such as GST in Canada and VAT overseas
- Setup charges, such as to prepare the plotter, set up the VCR
- Material charges, such as media, pens, and video tape
- Delivery charges, such as packaging fee and COD charges

Contract Considerations

When having drawings digitized, specify to the service bureau the file format (AutoCAD, MicroStation, Cadvance, CadKey, et cetera) that you want the drawing returned in. Check on the media too: size of diskette or format of tape.

When having drawings plotted, specify to the service bureau how you want the plot performed. Ensure ahead of time that the bureau's computers can read the drawing files you supply. Include a form that lists the layers and colors matched with pen widths and sizes. Specify the output media (paper, mylar), pen type, and media size.

Ask about the bureau's satisfaction guarantee, including items such as:

- Delivery time
- Cancellation notice
- Refund policy
- Dispute resolution

Also inquire into the bureau's discount schedule. Most bureaus charge less per item for large amounts of work coming their way. For example, a service bureau in California gives every fourth student from the same firm free tuition.

Summary

In this chapter, you learned about working with a service bureau, the services they provide, and their pricing. In the next chapter, you learn about working with clients, particularly in the area of translation drawings.

CHAPTER 12

Data Exchange

Perhaps your firm has two or more CAD packages in-house or deals with a client who uses a different package. You face the twin problems of translating drawing files and translating the drawings accurately. This chapter describes how to exchange drawings using DXF and DWG files between AutoCAD, MicroStation, Cadvance, Generic CADD, AutoSketch, DesignCAD, and Drafix Windows CAD.

Proprietary Drawing Files

Each CAD package uses a proprietary format to store its drawing information in files on disk. None of the formats is compatible with any other. AutoCAD does not read a MicroStation DGN file or a Cadvance VWF file or a Generic CADD 6.x GCD file. (Even though Generic CAD 5.0 stores drawings with a file extension of "DWG," the file format is incompatible with AutoCAD's DWG format.)

Intergraph and American Small Business Computer are exceptions. Intergraph documents most (not all: (newly added objects, such as shared cells, are not documented) of MicroStation's DGN file format and American Small Business Computer documents DesignCAD's DW2 format in the user manual. This makes it easier for third-party developers to write direct translators.

Executive Summary

While it is preferable that drawing files not be translated from one CAD package to another, sometimes it is unavoidable. The challenge is to translate the drawing accurately. The five prominent file formats used in translating drawings are:

- **DXF** is the *de facto* standard among PC-based software, including CAD, desktop publishing, and analysis program

- **IGES** is the official standard but is mostly used among host-based and workstation CAD packages

- **HPGL** is the translation format of last resort. All CAD packages produce it and many read it, but some information is lost

- **WMF** is the standard for exchanging vector information under Windows, although some information is lost, as with HPGL

- **DWG** is rapidly gaining acceptance as the translation format of choice since it eliminates the need for intermediate files

If you are faced with drawing files that must be translated into another format, there are three approaches you can take:

1. Perform the translation in-house. Your firm has control over the quality and delivery of translated drawings. But, your firm must purchase the translation software and, possibly, the other CAD package; you must provide the manpower to perform the translation and clean-up work.

2. Ask the client to perform the translation. The responsibility for a clean translation is transferred to the client. But, the client may bill your firm the cost; you and your client may not agree on what constitutes a clean translation.

3. Have a service bureau perform the translation. The translation work does not tie up your staff; the contract price maintains the budget. But, frequent use of a service bureau may be more expensive than performing the work in-house; the service bureau may be inexperienced at translation work. ■

Drawing Exchange File Formats

Since most CAD vendors keep the format of their binary drawing files secret, they provide alternate methods for importing and exporting drawings to and from outside sources. The five most common are:

- **DXF** (short for drawing exchange file) has become the *de facto* standard for exchanging drawings between PC-based CAD systems and other graphics programs

- **IGES** (short for initial graphics exchange specification) is more commonly used between workstation and host-based CAD systems

- **HPGL** (short for Hewlett-Packard graphics language) is written by all CAD packages; some CAD packages read the format

- **WMF** (short for Windows metafile format) is the standard for exchanging vector images among Windows applications

- **DWG** (short for drawing) is Autodesk's file format for AutoCAD drawings, which a number of outside programmers have cracked

In the following sections, we examine these five exchange formats in greater detail.

DXF

The DXF format was developed by Autodesk in the early 1980's for third-party developers to extract data from AutoCAD drawings for use by other programs. The DXF format has since become the *de facto* exchange format for almost all desktop CAD packages, something Autodesk did not intend. Many non-CAD graphics programs, such as desktop publishing, drawing programs and finite element analysis programs, also read DXF files. It takes two steps to convert a drawing file from one CAD format to another via DXF: (1) convert the CAD drawing file into DXF format, and (2) convert the DXF file into the other CAD format.

- **AutoCAD** has its DXF translator built into the Drawing Editor

- **MicroStation**'s DXF translator is a pair of MDL (MicroStation development language) apps called DxfIn.Ma and DxfOut.Ma

- **Cadvance**'s translator is a Windows DLL (dynamic link library) called Dxf.Dll

- **CadKey's** DXF translator is a CDE (CadKey dynamic extension) called DxfXl.Cde.

- **Drafix CAD Windows** uses a built-in translator or performs batch translations with a stand-alone program called DxfPort.Exe

- **Generic CADD's** translator is an stand-alone program called ACon.Exe

- **DesignCAD's** DXF translator is a stand-alone program named DcFiles.Exe

- **AutoSketch's** DXF translator is built into the core program

In all cases, the translator is called from the CAD program's command or menu area.

IGES

The IGES format was developed by a committee representing CAD vendors and government agencies to be an industry-standard neutral translation format. Since it was developed by committee, IGES is designed to handle almost any entity from any CAD package; unlike DXF, IGES is not limited to entities defined in AutoCAD. Like DXF, you go through two steps to translate between two CAD systems: (1) convert the CAD drawing file into IGES format, and (2) convert the IGES file into the other CAD format.

IGES is primarily used on workstation and mainframe CAD systems. It found disfavor among desktop CAD systems due to the extremely large files it produces. An IGES file is approximately three times larger than the equivalent DXF file and six times larger than the source file.

- **AutoCAD** has an out-of-date IGES translator built into the Drawing Editor; the up-to-date IGES translator is optional extra-cost software

- **MicroStation** 5 includes the IGES translator; for version 4, the translator is an optional extra-cost software package

- **Cadvance** does not supply an IGES translator

- **CadKey** includes a stand-alone IGES translator and editor called IG2C.Exe

- **Drafix CAD Windows** has an IGES translator built into the core program

- **Generic CAD** once sold an IGES translator as an optional extra-cost program but no longer does due to lack of demand

- **DesignCAD** includes an IGES translator as part of the stand-alone program named DcFiles.Exe

- **AutoSketch** does not support IGES

HPGL

Since all CAD software's ultimate purpose is to produce a drawing with a plotter, they all output HPGL-format files, the *de facto* standard among pen plotters. The data sent to a plotter is in vector format; thus it is easy to read the HPGL file and convert the plotter line information back into CAD vectors. There are two steps to translating a drawing using HPGL format: (1) plot the drawing to a file on disk in HPGL format; do not use the HPGL/2 format, and (2) read the HPGL file into the CAD program.

While color and text is retained, layer and attribute information is lost. Thus, HPGL should be used as a translator of last resort. AutoCAD, AutoSketch, Cadvance, and CadKey write but do not read HPGL files; Generic CADD, Drafix CAD Windows, and DesignCAD read HPGL files.

WMF

Most Windows applications accept images in the WMF format via the Clipboard. This is the quickest way to exchange drawings since there is no time taken to write the file to disk; some applications, such as AutoCAD, read and write WMF files to and from disk. The WMF format (also called the "picture" format by Microsoft) supports both raster and vector elements. There are three steps to exchanging drawings via WMF: (1) copy image to the Windows Clipboard with the **Edit | Copy** menu selection, (2) if necessary, select the WMF format (called "picture") in the Clipboard Viewer with the **View | Picture** menu pick, and (3) paste the image from the Clipboard with the **Edit | Paste** menu selection.

Like HPGL format, WMF looses layer and attribute information; colors, line widths, area fills, text, and vectors are retained. The Windows versions of AutoCAD, MicroStation, Cadvance, and Drafix CAD Windows read and write WMF format; AutoSketch only exports WMF. Generic CADD, CadKey, DesignCAD, and the DOS version of AutoCAD do not work with WMF files.

DWG

Since AutoCAD has a 60% share of some CAD markets, competitors have begun to include DWG as a option for importing drawings. While Autodesk does not publish its DWG file format, third-party developers have cracked the binary code. The advantage to DWG is that it removes the extra step of translating an AutoCAD drawing into a neutral format. There is a single step involved in translation: read the DWG file into the other CAD package.

In addition, MicroStation, CadKey, Cadvance, and Generic CADD make reading a DWG more transparent than reading DXF or IGES. The drawback is that, like DXF, the DWG format is limited to entities defined in AutoCAD. Thus a MicroStation ellipse does not translate to Cadvance as an ellipse, since DWG does not define an ellipse entity.

AutoCAD gives you no control over the import and export of DXF and DWG files. Fortunately, most of the "other" CAD packages' translators are equipped to cross-reference inconsistencies. Even with cross-referencing, expect that your drafters will have to do some editing after the translation. MicroStation, CadKey, Cadvance, Generic CADD have two-way DWG translation; AutoSketch is due to have the feature added with its next release; Drafix CAD Windows and DesignCAD do not read or write DWG format.

The Horror of Drawing Translation

While CAD vendors have made it a straight-forward matter to convert a drawing, the conversion process creates uncertainty over drawing database integrity. A very large drawing can loose entities that go unnoticed; even when no errors occur, you need to check every translated drawing for possible errors.

It's easy to make the incorrect assumption that all CAD packages have the same basic entities, such as point, line, arc, and circle. In fact, it is surprising how much they differ. Here are two examples.

Example: AutoCAD does not have a true ellipse entity as found in most other CAD packages; instead, it approximates an ellipse with polyline arcs.

Example: MicroStation doesn't have points or circles; instead, a point is approximated by a zero-length line, while a circle is drawn as a 90-degree ellipse.

To handle differences in entity types, drawing translators perform an approximate conversion that may result in inaccuracies.

Unique Entities

Most CAD packages have one or more unique entities. No other CAD package has an entity identical to AutoCAD's variable-width polyarc and no CAD package accurately duplicates it. In turn, AutoCAD does not have "weight" (a relative line width), which other CAD packages support. For unique entities, drawing translators perform either an approximate conversion or make no attempt to translate the entity. The Table shows how three CAD packages deal with AutoCAD's entities in the CAD package's own dialect. The ellipses "..." indicates the entity is not translated and is not displayed.

Translated Entities

AutoCAD 12	MicroStation 4	Cadvance 5	Generic CADD 6
Arc	Arc	Arc	Arc
Associative dimension	Exploded into lines, text and shapes	Group	Exploded dimension
Attribute	...	Text	Attribute
Block	Cell or exploded into linestrings	Symbol	Component
Circle	Round ellipse	Circle	Circle
Entity Handles	...	...	...

AutoCAD 12	MicroStation 4	Cadvance 5	Generic CADD 6
Extended entity data	...	...	...
Line	Line	Lineset	Line
Paper space	Sheet file	...	...
Point	Zero-length line	Invisible node	Point
Polyline	Complex chain	Lineset	Line
Polyarc	Complex chain	Lineset or arc	Line
Donut	Cell, if filled;	.	.
Ellipse	.	.	.
Polygon	.	.	.
Spline	.	.	.
Wide	.	.	.
Variable width	.	.	.
Region 2D model	3D drawing	2D drawing	2D drawing
Shapes	...	...	...
Solid 3D model	3D drawing	3D drawing	2D drawing
Solid	Primary shape	Lineset	Filled line
Text	Text	Text	Text line
Trace	Primary shape	Lineset	Filled line
Viewport	Reference file attachment	...	...
Xref	Reference file attachment	...	...
3D face	Primary shape	Face 3D	2D line
3D poly	Line string or b-spline curve	Lineset 3D	2D line
3D mesh	3D b-spline surface	Face 3D	2D line

Entity Limits

CAD packages differ in their limits on objects, such as layer names and line types. AutoCAD tends to be unlimited in its capacity, since layers, blocks and linetypes can have names with up to 31 characters. Effectively, you can insert an unlimited number of layers, blocks and linetypes into an AutoCAD drawing. Other CAD packages tend to have limitations, as the shown in the Table:

CAD Entity Limits

CAD Object	AutoCAD 12	MicroStation 4	Cadvance 5	Generic CADD 6
Database	64-bit real	32-bit integer	32-bit integer	32-bit integer
Max. colors	255	255	15	255
Max. linetypes	Unlimited	8	8	10
Max. styles	Unlimited	128	5	Unlimited
Max. layers	Unlimited	63	255	255
Layer names	31 characters	6 characters	15 characters	8 characters
Hatch names	31 characters	6 characters	12 characters	8 characters
Block names	31 characters	6 characters	5 characters	8 characters

Errors occur when a drawing exceeds the limits of the target CAD system. During translation, layer and block names are truncated, entities are combined onto layers, linetypes may not match and text looses font definitions.

CAD Database Accuracy

AutoCAD is one of the few to use 64-bit real numbers to define entity vectors, while most other CAD packages use 32-bit integers or real numbers. You may be used to drawing at any size in AutoCAD; in MicroStation, you need to first defined the minimum distance increment, which then defines the largest drawing area. While integers limit the size of very large projects, they are more accurate than real numbers for measuring very small differences.

Text Entities

Text is a special case, since it is the only entity whose shape depends on its definition. Tests show that CAD packages tend to translate only four

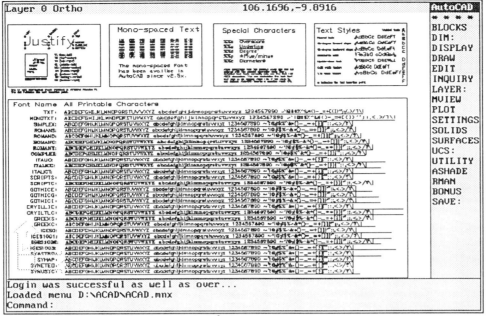

Figure 1: *This AutoCAD drawing tests text fonts.*

or five text styles defined in an AutoCAD drawing. The remaining styles are either assigned the CAD system's basic font, or else ignored—the text does not display.

When text is not represented by the precisely identical font definition, the text changes it size. A paragraph of text can only take up the exact same area when the character height, character width, intercharacter width and line spacing match exactly. Otherwise, you end up with text squeezed closely together (hard to read) or spaced too far apart (it overwrites other parts of the drawing). For true compatibility, the CAD vendors must include text fonts that match exactly. Only Generic CADD provides a set of fonts that exactly match AutoCAD's; MicroStation and Generic CADD include an AutoCAD font file converter.

Even when text fonts match exactly, there are further problems of justification, text styles, and special characters. The good news is that CAD packages deal with most of AutoCAD's 14 justification modes; only the aligned and fit modes tend to cause problems. The bad news is that CAD packages can't deal with most of AutoCAD eight text styles (such as forward sloping and backwards) or seven special characters (those prefixed by %%). The Table shows how three CAD packages react to AutoCAD's treatment of text. The ellipses "..." means text is translated as right-justified.

Text Translation

AutoCAD 12	MicroStation 4	Cadvance 5	Generic CADD 6
Justification Modes	Does not translate Aligned and Fit modes	Does not translate Fit and two-letter modes	Does not translate Aligned and Fit modes
Text Styles			
Forward slope	...	Correct	Correct
Backward slope	...	...	Correct
Backward text	...	...	Correct
Upside-down text	...	...	Wrong justification
Vertical text	...	...	Correct
Width factor	Correct	Correct	Correct
Special Characters			
%%%	Appears as %%%	Appears as %%%	Appears as %%%
%%c	Appears as %	Appears as %%c	Appears as c
%%d	Correct	Appears as %%d	Correct
%%nnn	...	Appears as %%nnn	...
%%o	Appears as %	Appears as %%o	Appears as o
%%p	Appears as *	Appears as %%p	Correct
%%u	Appears as %	Appears as %%u	Appears as u

DXF Via AutoCAD

AutoCAD has its DXF translator built into the Drawing Editor. The DxfOut command converts the drawing currently loaded into the Drawing Editor into DXF format, as follows:

```
Command: dxfout
File name <default>: <Enter>
Enter decimal places of accuracy (0 to 16)/Entities/
   Binary <6>: <Enter>
```

Since Autodesk defines the DXF standard, AutoCAD gives you little control over the translation results. Your choices are limited to:

- **Decimal places of accuracy:** the default of six decimal places is adequate for all but the most exacting translation requirements.

- **ASCII or binary format:** always use ASCII format (the default) unless you know that the receiving program read binary DXF format. The binary file is always written to 16 decimal places.

- **Complete drawing or entities only:** when translating a drawing, you have the option of exporting the entire drawing (the default) or just a portion of the drawing.

With each new release of AutoCAD, Autodesk makes changes the DXF format to account for new features. AutoCAD reads DXF files created with almost any version of the DXF format, up to and including the current version. Since the DXF format changed radically with AutoCAD v2.0x, current versions of most programs do not read DXF files produced prior to v2.0. While AutoCAD Release 10 can read a Release 9-format DXF file, it cannot read a Release 11-format DXF file. Autodesk provides the DxfIx.Exe program with AutoCAD to perform downward translations.

To import a DXF file, you use AutoCAD's **DxfIn** command, as follows:

```
Command: dxfin
File name:
```

Once the file is successfully read in, AutoCAD automatically performs a Zoom All to display the entire drawing. You save the file in AutoCAD's DWG format with the Save command.

Although the DxfIn command presents you with no options, AutoCAD treats a DXF in two different ways:

1. If the DXF file is read into a new "empty" drawing, AutoCAD will load the entire DXF file. You create an empty drawing in AutoCAD Release 11 and earlier by including an equals (=) sign at the end of the new drawing's file name in the Main Menu, as follows:

   ```
   Enter NAME of new drawing: filename=
   ```

   With AutoCAD Release 12, you create an empty drawing by clicking on **File | Open**, then clicking the check box next to **No Prototype**.

2. If the DXF file is read into a drawing that is not empty, AutoCAD displays the message, "Not a new drawing -- only ENTITIES section will be input" and loads only the entities from the DXF

file. Information stored in the DXF file about layer names, block definitions, text styles, et cetera is ignored.

The AutoCAD translator is very touchy. If it come across a single error in the DXF file, it aborts the loading process with the error message, "*Invalid* An error was discovered while loading a drawing." AutoCAD then discards any data already loaded into the current drawing. In some cases, AutoCAD will report the line number that it found the error. The line can be edited by someone knowledgeable about DXF files.

DXF Via Third-Party Software

Many firms that don't normally use AutoCAD still maintain one copy in house. It performs batched DXF translations and is used to verify the translation.

If your firm's translation needs are predictable, you may want to employ one of several third-party programs that read AutoCAD DWG files and produce DXF files. That saves you the expense of purchasing AutoCAD. For example, Drawing Librarian (SoftSource) reads AutoCAD DWG files and translates them into DXF formats compatible with Release 9, 10, and 11. The program also views DWG and DXF files to verify the translation.

The Test Grid

Looking for translation mistakes is time-consuming and inaccurate. How do you ensure a drawing translation is accurate? Create a grid of boxes with the CAD program, placing an entity and label in each box. The manual usually has a list of the CAD program's basic entities. Figure 2 shows a sample grid created with Auto-CAD Release 11 (missing is the viewport entity, visible only in paper space). You may

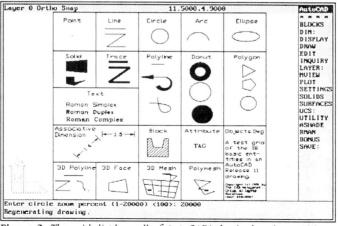

Figure 2: *The grid displays all of AutoCAD's basic drawing entities, with the exception of viewports.*

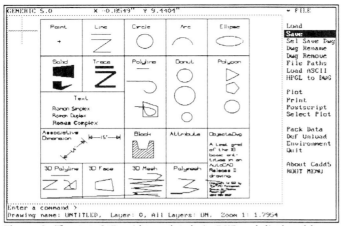

Figure 3: *The AutoCAD grid translated via DXF and displayed by Generic CADD.*

want to include special constructions your firm uses, such as nested blocks and unusually configured text fonts. Translate the file and bring it up in the other CAD package. At this point, you look for obvious abnormalities. Figure 3 shows the result of translating the AutoCAD grid into Generic CADD 5.0 with AutoConvert. The DXF translator had difficulties with AutoCAD's polyline, 3D mesh and polymesh entities, while the attribute is missing entirely. The newer version of Generic CADD translate AutoCAD drawings more accurately. Plot both drawings at the same scale on mylar or some other transparent media. By overlaying the two plots, you pinpoint less obvious inaccuracies, such as text spacing.

Summary

Due to inherent incompatibilities between all CAD packages (even from the same vendor), it is best not to translate drawings. However, if you have no choice, proceed with caution recognizing that drawing accuracy may be compromised during translation.

CHAPTER 13

Networking CAD

Networking lets CAD workstations share drawing files and hardware devices, such as backups, plotters, CD-ROM drives, and hard disk space. This chapter looks at the benefits and problems of networking a CAD system.

What is a Network?

When you work with CAD on more than a single workstation you begin to network. You find that the CAD operators need to share files. Your accountant finds it's not cost effective to attach a plotter to every workstation. Sharing resources, such as files and peripherals, is what *networking* is all about.

Since installing a network is a significant cost, firms tend to ignore it initially. Instead, they employ a type of networking commonly known as SneakerNet, which involves moving files between computers by carrying a diskette. To solve the problem of one plotter servicing more than one CAD station, you dedicated a computer to running the plotter; files are carried on diskette between the CAD stations and the plotter station.

After some months (or years) of copying files to and from diskettes, the process becomes tedious and firms begin to investigate networks. Until recently, networks were a gamble to install: there were more competing standards than colors in the rainbow, coupled with the high cost of NICs (network interface cards).

Today the situation is very much different. Standards have settled down and prices have plummeted.

> ## Executive Summary
>
> A network consists of three parts:
>
> 1. **Network operating system:** redirects file and and peripheral requests from the local computer to the network. Windows, Windows NT, Macintosh, and Unix systems all include a network operating system of greater or lessor capability.
>
> 2. **Network interface card:** connects the local computer with the network cabling. While several standards exist, only use EtherNet.
>
> 3. **Network cabling:** wiring that connects computers together. Fiber optic is most expensive; twisted pair is least expensive; thin coaxial is prices inbetween and the most common.
>
> There are two kinds of network styles:
>
> 1. **Central file server:** all computers are connected to one central computer with an enormous hard drive.
>
> 2. **Peer-to-peer:** all computers are connected to each other. ■

The Components of a Network

A network consists of three primary parts:

- Network operating system

- Network interface card

- Network cable

Network Operating System

DOS and Windows used to be network-hostile. Today, Microsoft is integrating network functions into the core of the operating systems. MS-DOS v5 and higher has the DOS Connection option that lets any DOS-based computer network with Windows-based computers.

Windows has the Windows for Workgroups upgrade that integrates networking. WFW includes software useful for networking: group scheduler, electronic mail, network monitor, and a chat facility. Windows NT has inherent network capabilities and comes in two flavors: "basic" NT and Advanced Server. Windows NT Advanced Server is meant for a central file server network.

Network Interface Card

Most NICs are under $200 with some models now less than $100. Some computers come with the NIC circuitry built into the motherboard. There are even external NICs that connect to the computer's parallel port—meant for notebook computers but let users install the network hardware without even opening the computer. There are a number of different NIC standards available: purchase only those NICs compatible with the EtherNet standard.

Network Cable

The third component of a basic network is the cabling. While fiber optic cabling is glitzy, it is expensive. The two most common wire-based cables are "twisted pair," which is a strange name for ordinary telephone-style cabling, and "thin coax," which is a thinner version of the wiring used for cablevision (called "thick coax").

Two Styles of Networking

You connect workstations together with a network in two styles, called "central file server" and "peer-to-peer."

Central File Server

In the central file server scheme of things, all workstations are connected back to a computer whose primary attribute is a monster-size hard disk—at least one gigabyte (1,024 megabytes) in size. The central file server holds all software and data files.

When an operator wants to start a CAD program at their workstation, a batch file loads the CAD software along the network from the central file server. The drawing and other support files all reside on the central file server. Only temporary files reside on the workstation. In addition, all significant peripherals are connected to the central file server, such as the tape backup units, the plotters, and high-speed printers.

The advantages and disadvantages to a central file server are:

- Installing a software update involves a single installation

- A central file server network software is usually more feature-rich and faster than the peer-to-peer network software

- If the central file server crashes, the entire network is debilitated

- Higher cost than a peer-to-peer network

Peer-to-Peer

In a peer-to-peer network, all workstations are connected to each other. The network software limits access to specific subdirectories on other people's workstations. Each workstation might be connected with a specific peripheral but all workstations can access it.

The advantages and disadvantages to a peer-to-peer network are:

- Lower cost than a central file server

- If one computer crashes, the network continues to operate

- However, the peripheral attached to the crashed computer is unavailable

Windows NT (basic version) and Windows for Workgroups are meant for peer-to-peer networks.

Multiple LANs

It is possible for both styles of networks to be connected. Multiple LANs (local area networks) are connected with each other via "hubs." This is referred to as a WAN (wide area network).

With appropriate hardware and software in place, any workstation on a network can access data located in any DOS-based, Macintosh, Unix-based, mini-, and mainframe computer. Using a combination of LANs and high-speed, dedicated network connections, it is possible for a global corporation to have all its offices and computers networked around the world.

The Complexity of Networks

With the brief description of networking so far, you begin to sense their complexity. On the one hand, networks give your staff the freedom to share data and peripherals, But, as with any freedom, limits must be applied.

Traffic Copy Functions

The network must also act as a traffic cop. Consider these situations:

- Access to the workstation's subdirectories must be restricted to read-only access, read-and-append access, full read-and-write access, or no access at all.

- Drawings sent to the plotters and printers must be queued.

- Data must be backed up from all workstations onto the central file server or to a tape backup on a daily basis. The backup must occur in a short enough time so as not to interfere with the normal use of the network.

- Platform differences means being aware of the differences between computers. The most common example is filename. DOS allows eight-plus-three characters, Mac allows 31 characters, OS/2 HPFS and NT NTFS allow 256 characters, and Unix allows a 16-character name that includes an optional extension.

The LAN Manager

Most firms find they need a dedicated LAN manager, who may well be the same person as the CAD manager. It is the LAN manager's job to:

- Set the rules for which users get to access which files and subdirectoris of data

- Get workstations connected to the network

- Ensure the backup tapes are rotated in order

- Get the network back on-line after it crashes

Additional Benefits of a Network

Once the network is installed, you will find it is more than just a way to share files and peripherals. Network software is bundled with increasingly useful software. Here are some examples:

- Electronic mail is a silent, fast, environmentally-friendly way to communicate office memos and notes to staff (see Figure 1)

- Access to external email, such as Internet, lets you communicate with email-savvy clients and branch offices via electronic mail

- Network fax boards let every user to send and receive faxes via the network, with a connection to the office PBX box

- Scheduling software lets managers poll staff to see when meetings are possible (see Figure 2)

- With all files on-line, managers can view a project's progress with drawing viewer software

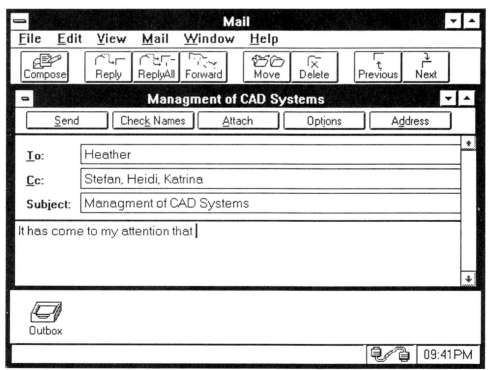

Figure 1: *Electronic mail over a network lets users communicate instantly with each other.*

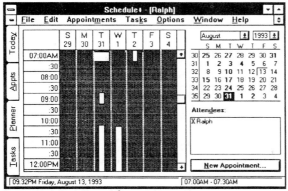

Figure 2: *Scheduling software that operates over a network makes it easier to schedule meetings.*

- With appropriate front-end software, a network allows you to implement an electronic drawing check-in/out procedure

- CAD operators can externally reference drawings, no matter on whose workstation the drawings are located

- Plots can be queued up in the network, instead of people queuing at the plotter. Some plot queue (or "spooler") software lets the LAN manager change the priority of plots in the queue

- When staff are on the road or working from home, they can hook up a modem to their computer and do a "remote dial-in" that lets them access all resources on the network as if they were in the office

- Backups become reality since the user is no longer responsible; instead, the network is programmed to automatically backup the entire system during the night

Features of a Network-Friendly CAD Package

While any software can run off a network, CAD software needs special features because of its "groupware" nature. Some features to look for are:

- **File Locking.** The CAD package must be able to lock its drawing files to prevent a second user from changing a design already being edited by someone else.

- **Shared Licenses.** Network-friendly software lets you install a single copy, then allows operators to open multiple copies. A "floating network license" scheme allows as many copies be opened as your firm purchased license for.

- **Plot to Queue.** Instead of plotting to the serial port, a network-compatible CAD system can be directed to plot to a special spooler name. This lets the network take care of queuing and redirecting plot files to the appropriate plotter.

- **Redirection of Temporary Files.** CAD software often opens a large number of temporary files, which cannot be confused between multiple operators. Network-friendly CAD software lets each workstation redirect the temporary files to the local subdirectory.

- **Cross-platform Compatibility.** Because networks often involve disparate computers (PCs, Macs, and Unix machines), the CAD software's data files should be platform-independent. That means that a drawing file drawn on a Mac should not need translation in order to be loaded on the PC version of the software.

Figure 3: *Networks implement password protection to prevent unauthorized use of programs and data.*

- **Password Protection.** To protect CAD drawings from being edited by unauthorized persons or disgruntled staff, the CAD software should be protected by a password, refusing to load if the wrong password is entered (see Figure 3).

- **Network Node Name.** By assigning a unique name to each workstation, the source and ownership of files can be easily determined.

- **Externally Referenced Drawings.** This lets a CAD operator view drawings stored on another computer as a backdrop to the operator's current drawing—very important for multi-discipline projects.

Summary

This chapter provided an introduction to the benefits and concerns in connecting CAD workstations with a network. In the next chapter, you learn about techniques that optimize CAD operations.

CHAPTER 14

Maximizing CAD Efficiency

Running CAD software straight out of the box proves to be an exercise in frustration. With the exception of a few, CAD packages are generalized drafting and design engines meant to be customized to your discipline. This chapter describes the seven areas where most CAD packages can be customized.

Why Customize CAD?

The manager of the roof truss division of a large forest products company was on the phone: "What kind of hardware do we need to make our CAD system run faster?"

I replied that there are many ways to make the hardware faster but I thought to ask him, "Why?"

He explained that their CAD operator was producing drawings slower by CAD than by hand, even though she had six months training at a local college. When he had tried to find out why drafting production had regressed, her explanation was that the computer was too slow.

On a hunch, I asked him: "Does she use blocks?"
"No."
"Has she customized the menus?"
"No."
"Has she programmed repetitive routines?"
"No."

The solution was simple. "The fastest hardware in the world won't speed up your drafter if she hasn't customized the CAD software for drawing roof trusses," I told him.

Executive Summary

Most CAD packages must be customized to work most efficiently in your office. This chapter describes seven areas where most CAD packages are customizable:

- Parametric symbol libraries

- Screen and tablet menus

- Icon menus and toolbars

- Macros and scripts

- Built-in programming languages and links to external languages

- Database links

- Under Windows: dynamic data exchange, object linking and embedding, and Clipboard support.

The chapter includes lists showing the customizability of eight CAD packages: MicroStation, AutoCAD, CadKey, Cadvance, Drafix CAD, Generic CADD, DesignCAD, and AutoSketch. In addition, the following techniques can help speed up any software:

- Max out on memory

- Move that memory up high

- Cache that disk

- Spool that plot

- Co-process those graphics

- Process that display-list ■

After spending $3,000 or more on a CAD package, it comes as a shock to many users that the CAD software isn't very useful straight out-of-the-box. As the story illustrates, throwing a lot of hardware at slow production speed is expensive and less effective than applying software solutions. The clever application of software can help a drafter produce drawings faster on an optimized 386 than on an unoptimized 586/Pentium.

The previous chapters of this book have lead you through a number of steps that have already helped optimize your CAD system. Creating a set of prototype drawings, standardizing on consistent standards, working with a service bureau, and networking your firm's computers lets your staff get their jobs done faster.

In this chapter, we focus on the nitty-gritty features provided by the CAD software for optimizing it to your discipline. CAD software can be customized in a number of ways; the most common are:

- Parametric symbol libraries

- Screen and tablet menus, icon menus and toolbars

- Macros and scripts

- Built-in programming languages and links to external languages

- Database links

- Under Windows: dynamic data exchange, object linking and embedding, and Clipboard support

In the following sections, we look at how each of the eight previously-reviewed CAD packages implement customization.

Parametric Symbol Libraries

Earlier in the book we looked at how to create symbols and symbol libraries. It's clear that placing a symbol with a couple of menu picks is much faster than redrawing each symbol over again.

Going one step further, parametric symbols store more efficiently on the hard disk. Instead of one symbol file for every type of bolt, one parametric file holds the design parameters for all possible bolt designs. The CAD operator selects the type of bolt and the CAD's programming facility creates the symbol on-the-fly.

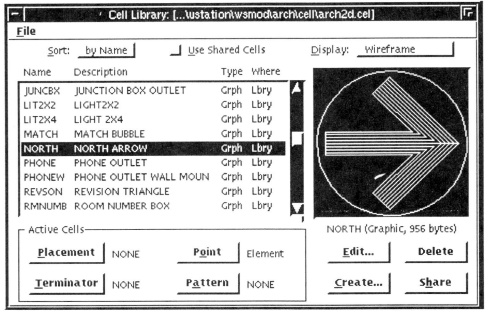

Figure 1: *Several CAD packages (MicroStation shown here) organizes symbols ("cells") into libraries and previews the symbol before placement.*

Here is a brief listing of the symbol capabilities in eight CAD packages:

- **MicroStation:** symbols, symbol libraries, symbol preview, and parametric symbols.

- **AutoCAD:** symbols and parametric symbols.

- **CadKey:** symbols and parametric symbols.

- **Cadvance:** symbols and parametric symbols.

- **Drafix CAD:** symbols, symbol libraries, symbol preview, and parametric symbols.

- **Generic CADD:** symbols.

- **DesignCAD:** symbols.

- **AutoSketch:** symbols and symbol preview.

Screen and Tablet Menus

From the earliest days of CAD software, users could customize the menus. The menus are used to display the available commands, group commands together logically, and give easy access to macros. CAD systems typically have two kinds of menus: (1) screen, and (2) tablet.

The screen menu is traditionally on the right or left edge of the screen (called the "side-screen" menu); more modern implementations place it along the top edge of the screen (called the "pop-down" menu).

The tablet menu consists of an array of boxes on the digitizing tablet. Since the tablet has a large area, it is possible to place almost all of a CAD system's command in plain view.

Some CAD systems have menus available in other locations, for example on pointing device buttons and on-screen menus that pop-up at the cursor location.

Under Windows, icon menus and toolbars have become popular. The icon menus typically float on the desktop, with icons explaining the function of the command or macro. The toolbar lines the top edge of the screen, typically under the menu bar. The toolbar usually provides a

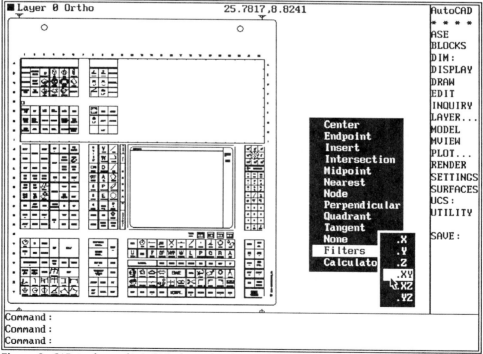

Figure 2: *CAD packages (AutoCAD shown here) provide a variety of menus, including tablet, side-screen, pop-down, and pop-up menus.*

shortcut access to commands hidden in the nested menu structure.

Here is a brief listing of the menu capabilities in eight CAD packages:

- **MicroStation:** side screen menu (v3-compatible), pop-down menu, tablet menu, and floating icon menus.

- **AutoCAD:** side screen menu, pop-down menu, tablet menu, and pointing device buttons. Windows version has a toolbar and a floating icon menu.

- **CadKey:** side screen menu and tablet menu.

- **Cadvance:** pop-down menu, floating icon menus, and toolbar.

- **Drafix CAD:** pop-down menu and floating icon menu.

- **Generic CADD:** side screen menu, tablet menu, and pointing device buttons.

- **DesignCAD:** pop-down menu, tablet menu, and pointing device buttons.

- **AutoSketch:** pop-down menu, toolbar, and floating icon menus.

Macros and Scripts

Macros and scripts are the most rudimentary form of automating a software package. Macros imitate keystrokes and allow some additional commands, such as logic branching and prompting for user input. Script files usually only imitate keystrokes.

Here is a brief listing of the macro capabilities in eight CAD packages:

- **MicroStation:** *feature not available.*

- **AutoCAD:** macros are assigned to pointing device buttons or to menus; scripts are stored on disk. Windows version has macros assigned to a toolbar and floating icon menu.

- **CadKey:** macros are assigned to keystrokes and the tablet menu boxes; stored in libraries on disk.

- **Cadvance:** macros are assigned to icon buttons and are stored on

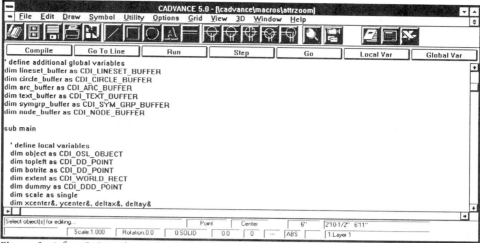

Figure 3: *A few CAD packages (Cadvance shown here) integrate a macro debugging facility.*

disk; includes a full macro programming environment.

- **Drafix CAD:** macros are written in DGL (Drafix Graphics Language) and are assigned to icon buttons, menus, or run with the @ symbol; includes a full macro programming environment.

- **Generic CADD:** macros are assigned to function keys, pointing device buttons, menus, or stored on disk (called "batch" files).

- **DesignCAD:** the macro feature records keystrokes; macros are assigned to the 12 [Alt] function keys.

- **AutoSketch:** the macro feature record keystrokes and can include a few commands; macros are given filenames.

Programming Languages

Many CAD packages include a powerful programming facility that lets you take full control of the CAD software. Usually, the programming facility is based on a common programming language, such as Basic, Pascal, and C. CAD software provide advanced programming in two ways: (1) the language is part of the package, and (2) programming links are provided to hook into an external language.

Here is a brief listing of the programming capabilities in eight CAD packages:

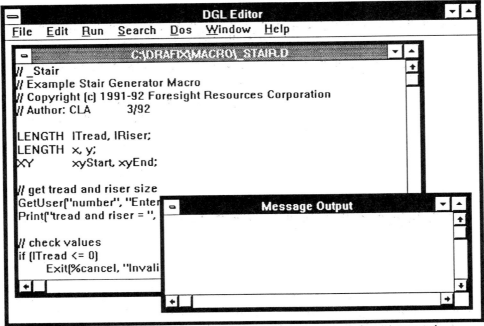

Figure 4: *Some CAD packages (Drafix CAD Windows shown here) include an integrated programming environment.*

- **MicroStation:** MicroCSL and UCM (user commands) are early assembler-like languages; MDL (MicroStation Development Language) is a C-like programming language that allows full control of MicroStation and full access to Windows DLLs; includes an integrated development environment.

- **AutoCAD:** AutoLISP is the built-in LISP-like programming language; ADS (AutoCAD Development System) is an interface to specific external programming languages, such as Microsoft C, Borland C++, and Visual Basic; both allow full control of AutoCAD.

- **CadKey:** CDE (CadKey Dynamic Extensions) is the C-based programming environment while CADL (CadKey Advanced Design Language) describes objects in the database.

- **Cadvance:** CBL (Cadvance Basic Language) is the Basic-based programming language, while CDI (Cadvance Development Interface) is a library of API calls with full access to DLLs; includes an integrated development environment.

- **Drafix CAD:** DGL (Drafix Graphics Language) is the Pascal and C-based programming language; includes an integrated development environment.

- **Generic CADD:** programming language is an extension of the macro language, based on no known programming language.

- **DesignCAD:** BasicCAD is a QuickBasic-like programming language that allows control of DesignCAD.

- **AutoSketch:** *feature not available.*

Database Links

A powerful feature of CAD software is its ability to link objects with database fields. Thus a line not only has length and direction, but it can represent a price, a material, and a catalog number. The database information is stored in either of two locations: (1) inside the drawing, and (2) external to the drawing.

When data is carried inside the drawing it is called "attribute" or "tag" data. Some CAD packages limit attribute data to symbols; others allow the data to be attached to any object. Carrying the data in the drawing is convenient because all data is in a single file; however, it is difficult to share the data with other drawings. Data inside the drawing is usually exported in CDF (comma-delimited format) or SDF (space-delimited format) for import into spreadsheets and database programs.

When data is external to the drawing, it is usually stored by a database program such as dBase, Paradox, and Oracle. Keeping the data external to the drawing makes it easier to share the data between drawings and applications; however, you have to set up links between the current drawing and the database file, which is intimidating to new users.

Here is a brief listing of the database capabilities in eight CAD packages:

- **MicroStation:** linkages to external databases (dBase and Oracle; in version 5, attributes are attached to symbols and entities.

- **AutoCAD:** linkages to external databases (dBase, Paradox, and ODBC under Windows) via ASE; attributes data is attached to symbols only.

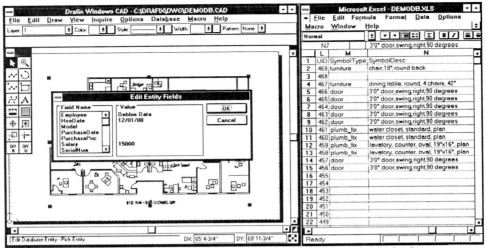

Figure 5: *Almost all CAD packages (Drafix CAD Windows shown here) export database information to spreadsheets (such as Excel), database programs, and word processors.*

- **CadKey:** no linkage to external databases; attributes are attached to symbols.

- **Cadvance:** linkage to almost any external database via Q+E; attributes are attached to symbols and entities.

- **Drafix CAD:** no linkage to external databases; attributes are attached to symbols and entities.

- **Generic CADD:** no linkage to external databases; attributes are attached to symbols only.

- **DesignCAD:** no linkage to external databases; attributes are attached to symbols only.

- **AutoSketch:** *feature not available.*

Under Windows

Windows provides three additional methods of customizing a CAD package: DDE (dynamic data exchange), OLE (object linking and embedding), and the Clipboard.

DDE lets up to 16 applications exchange data with each other in the background. In fact, AutoCAD for Windows uses DDE to communicate with its LISP and ADS modules. DDE also lets one application control another by sending macro commands. With the release of Windows for Workgroups, Network DDE lets an application send data to another application on a different computer via the network.

OLE lets you embed and link data from one application into another, without going through the programming required with DDE. For example, you can embed an AutoSketch drawing in a WordPerfect document. When you double-click on the drawing in WordPerfect, OLE automatically launches AutoSketch, letting you edit the drawing. If the drawing is linked, then any changes made to the drawing are automatically updated in the WordPerfect document.

The Clipboard is the mechanism that lets you cut, copy, and paste within an application and between applications. More important, it offers a form a data exchange. Thus, you can copy a symbol from a Cadvance drawing and paste it directly into an AutoCAD drawing—without bothering with the additional steps of translation the symbol via DWG or DXF. The Clipboard support any data format an application sends it: vector WMF, raster BMP, text, and original DGN/DWG/VWF/CAD/SKD formats.

Here is a brief listing of the Windows capabilities in eight CAD packages:

- **MicroStation:** Windows Connection fully support DDE, OLE, and the Clipboard.

- **AutoCAD:** Windows version supports DDE; is an OLE server but

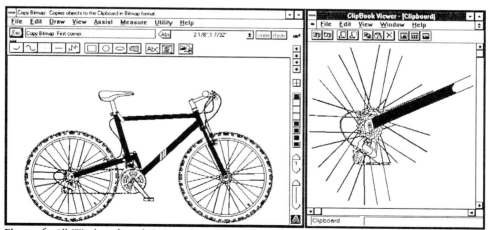

Figure 6: *All Windows-based CAD packages (AutoSketch shown here) support the Windows Clipboard in some way.*

not an OLE client; copy vectors in WMF and BMP formats to the Clipboard, paste WMF vectors and text from the Clipboard.

- **CadKey:** *features not available.*

- **Cadvance:** full support of DDE, OLE, and the Clipboard.

- **Drafix CAD:** full support of DDE, OLE, and the Clipboard.

- **Generic CADD:** *features not available.*

- **DesignCAD:** *features not available.*

- **AutoSketch:** does not support DDE; is an OLE server but not OLE client; copy and cut vectors in WMF format to the Clipboard but does not paste WMF from the Clipboard.

Optimizing CAD

So far in this chapter, we have focused on optimizing CAD by customizing the software. However, there are many other means of optimizing CAD that applies to software in general. Check that your workstations are employing these easy and low-cost speed-up techniques:

- **Max out on memory.** The more memory the workstation has, the faster memory-hungry software runs. Provide at least 8MB RAM for CAD packages, preferably 16MB.

- **Move that memory up high.** Memory managers move memory-resident device drivers and utility programs out of DOS' low memory (below 640KB) and into the upper memory (above 640KB) and high memory (above 1024KB) areas. Freeing up low memory lets DOS-based CAD packages run faster. Example products: Quarterdeck's QEMM-386, Qualitas' 386Max, and Microsoft's HiMem and Emm386.

- **Cache that disk.** Your workstations will experience a dramatic increase in speed by installing a disk cache. If memory is tight, size the cache at 256KB; if memory is plentiful, don't hesitate to set the cache size to 1MB or 2MB. Example products: Central Point Software's PC-Cache and Microsoft's SmartDrv.

- **Spool that plot.** The slowest stage in computer-aided drafting is the plotting stage. Install a plot spooler, which captures the plot output to a file on disk (a fairly quick process) then feeds the plot data to the plotter in the background, while your operators return to drafting. Example products: DOS's own Print command and Eclipse's Plump RX.

- **Co-process those graphics.** Graphics boards with coprocessor chips named S3, IIT, and 8514 display CAD graphics and Windows bitmaps faster than plain VGA graphics boards. Example products: CAD-specific boards from Artist Graphics, Metheus, and Vermont Microsystems.

- **Process that display-list.** AutoCAD and CadKey lend themselves to display-list processing, which is a form of display caching. Redraws, zooms and pans are 5 to 50 times faster with a display driver enhanced with display-list processing. Example products: Panacea's DLD, Vibrant's SoftEngine, and Nth Graphics' Nth Drive.

Summary

In this chapter you learned how to maximize the efficiency of CAD software. Thirteen different software approaches can make any CAD operator improve their output and make the computer appear to run faster.

Appendices

APPENDIX A

Gallery of CAD Software

Every CAD package has a unique set of strengths and weaknesses. With more than 200 CAD packages to choose from, it is not easy to pick one exactly right for your firm. For example, while clients might demand that your drawings be prepared with MicroStation, your internal projects might be better suited to Generic CADD.

Twenty-six CAD Packages

To give you an overview of the CAD software available today, the following pages present a representative mix of computer-aided design and drafting software. To present information on all CAD packages would require a separate book—tersely written (see Appendix B). While the earlier pages of this book presented the details of setting up eight CAD packages, here we present 26 that work with DOS- and Windows-based desktop computers.

All of the packages presented here are stand-alone CAD programs but some are better suited for conceptual design or as a front-end to other CAD packages, including Mannequin and Mannequin Designer, 3D Concepts, and Upfront.

The overview illustrates each package with a picture of the screen (lets you see the graphical user interface) and briefly lists its salient features, supported file formats, and forms of customization.

For more information, each listing includes the vendor's address, phone, and fax number. All CAD vendors have descriptive literature and some have demonstration diskettes available.

Anvil-1000MD
Interactive Design Consultants
Hawkeye Commercial Park, 12721 Wolf Road
Genesco IL 61254
(309) 944-8108   Fax (309) 944-6532

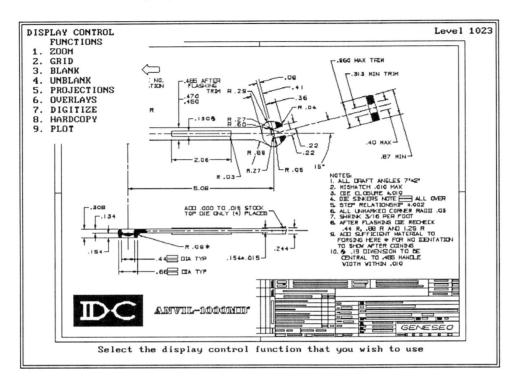

Features. Anvil-1000MD is a 2D-only mechanical drafting package that runs on a broad range of computers, from a DOS-based IBM XT through to Sun SPARCstations, IBM RS/6000, and Silicon Graphics. Largest drawing size: unlimited; layers: 1,024.

File Formats. Anvil's proprietary format is called DRW. It imports and exports drawings in DXF and IGES formats.

Customization. You can customize fonts, line weights, screen menus, tablet menus. Anvil can be programmed via keystroke macros and variable expressions for data input.

AutoCAD
Autodesk Inc
2320 Marinship Way
Sausalito CA 94965
(800) 445-5415

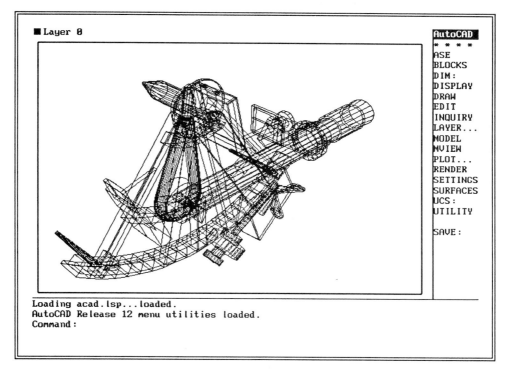

Features. AutoCAD is a 3D drafting and design package bundled with an SQL database link, PostScript fonts, full-color rendering, and a 2D region modeler. Largest drawing size: unlimited; layers: unlimited; international version uses a hardware lock.

File Formats. AutoCAD's proprietary format is called DWG. It imports and exports drawings in DXF, DXB, and IGES formats, plus several raster formats. Attributes are exported in DXF, SDF, and CDF formats.

Customization. You can customize fonts, linetypes, hatch patterns, screen and tablet menus, dialogue boxes, and digitizer buttons. AutoCAD can be programmed with macros, scripts, a LISP-like programming language, and has a link to C-language programs.

AutoCAD for Windows
Autodesk Inc
2320 Marinship Way
Sausalito CA 94965
(800) 445-5415

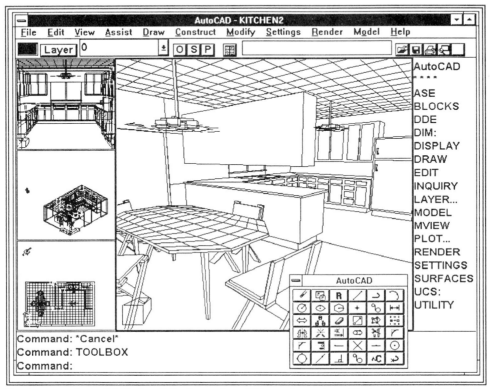

Features. AutoCAD for Windows is a 3D package bundled with an SQL link, PostScript fonts, full-color rendering, and a 2D region modeler. Largest drawing size: unlimited; layers: unlimited; international version uses a hardware lock.

File Formats. AutoCAD's proprietary format is called DWG. It imports and exports drawings in DXF, DXB, and IGES formats. Attributes are exported in DXF, SDF, and CDF formats. Windows Clipboard supports WMF, BMP, and import of text files.

Customization. You can customize fonts, linetypes, hatch patterns, screen and tablet menus, dialogue boxes, and digitizer buttons. AutoCAD can be programmed with macros, scripts, AutoLISP programming language, and link to C-language programs. AutoCAD is an OLE server and supports network DDE.

AutoSketch

Autodesk Retail Products

11911 North Creek Parkway South
Bothell WA 98011
(800) 228-3601 Fax (206) 483-6969

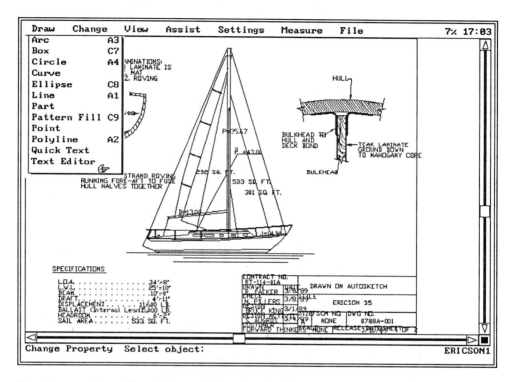

Features. AutoSketch is a 2D-only drafting package for DOS bundled with AutoCAD-compatible fonts and hatch patterns. Largest drawing size: 2MB; layers: 10.

File Formats. AutoSketch's proprietary format is called SKD. It imports and exports drawings in DXF and SLD.

Customization. You can customize fonts and hatch patterns, and write macros.

AutoSketch for Windows
Autodesk Retail Products
11911 North Creek Parkway South
Bothell WA 98011
(800) 228-3601   Fax (206) 483-6969

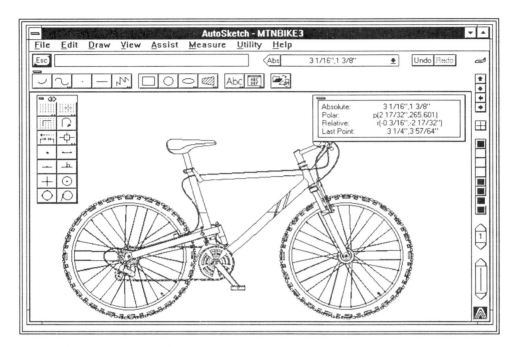

Features. AutoSketch for Windows is a 2D-only drafting package bundled with AutoCAD-compatible fonts, hatch patterns, and 500 symbols. Largest drawing size: unlimited; layers: 10.

File Formats. AutoSketch's proprietary format is called SKD. It imports and exports drawings in DXF and SLD. The Windows Clipboard imports and exports in SKD format, and exports in WMF and BMP format.

Customization. You can customize fonts, hatch patterns, and icon menus, and write macros. AutoSketch is an OLE server.

CadKey

CadKey, Inc.
4 Griffen Road North
Windsor CT 06095-1511
(203) 298-9999

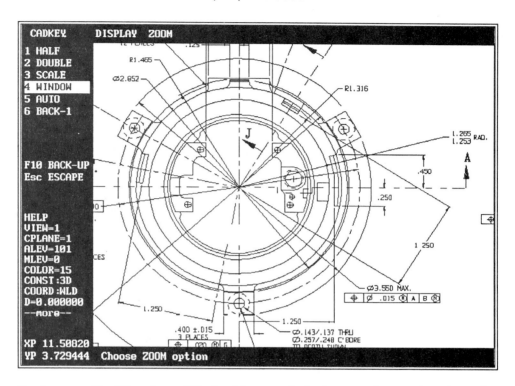

Features. CadKey is a 3D drafting and design package aimed at the mechanical market. It is bundled with an IGES translator, FastSurf, and integrated design analysis. Largest drawing size: unlimited; layers: 256; uses a hardware lock.

File Formats. CadKey's proprietary format is called PRT. It imports and exports drawings in DWG, DXF, and IGES formats. Attributes are not exported.

Customization. You can customize fonts, hatch patterns, screen and tablet menus. CadKey can be programmed with macros, CADL (CadKey advanced design language) and CDE (CadKey dynamic extensions).

CadMax TrueSurf
Cadmax Corp.
258 Village Square
Village of Cross Keys, Baltimore MD 21210
(410) 532-2803   Fax (410) 433-1192

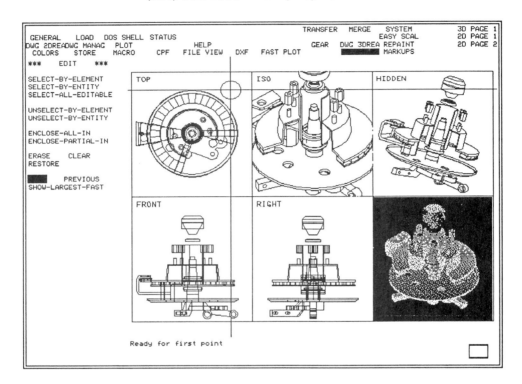

Features. CadMax TrueSurf is a 3D drafting and design package is based on surface modelling technology that produces hidden-line and shaded drawings in real-time primarily for the mechanical market. Largest drawing size: unlimited; layers: 255; uses a hardware lock.

File Formats. CadMax's proprietary format is called DWG but is not AutoCAD-compatible. It imports and exports drawings in DXF format. Attributes are exported in ASCII and SDF formats.

Customization. You can customize the screen and tablet menus. CadMax can be programmed with menus, macros and the CPF (CadMax programming facility) link to Pascal and C-language programs.

Cadvance for Windows
Isicad Inc
1920 West Corporate Way, PO Box 61022
Anaheim CA 92803-6122
(714) 533-8910   Fax (714) 533-8642

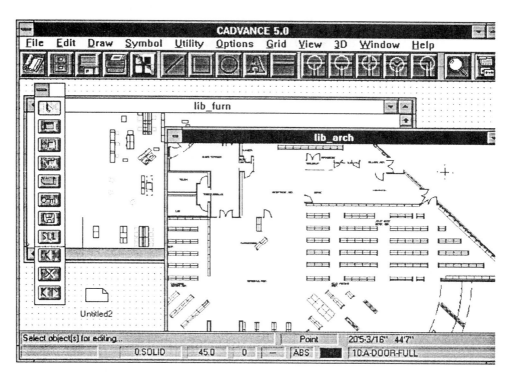

Features. Cadvance for Windows a 2D and 3D drafting package bundled with database links and workgroup support. Largest drawing size: unlimited; layers: 256.

File Formats. Cadvance's proprietary format is called VWF. It imports and exports drawings in DXF and DWG format, as well as via the Windows Clipboard in WMF, BMP, and text formats. Attributes are imported and exported via Q+E, which supports most database formats.

Customization. You can customize hatch patterns, line types, screen, tablet and icon menus, assign macros to icons and program with the Basic-like programming language. Cadvance is an OLE client and server and supports MAPI and bidirectional network DDE.

DesignCAD 2D
American Small Business Computers
One American Way
Pryor OK 74361
(918) 825-9367   Fax (918) 825-6359

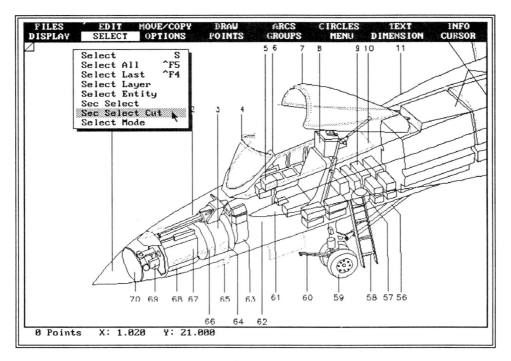

Features. DesignCAD 2D is a 2D-only drafting package for DOS bundled with 500 symbols and a bill of materials program. Largest drawing size: unlimited; layers: 256.

File Formats. DesignCAD's proprietary formats are called DC2 (ASCII format) and DW2 (binary format). It imports and exports drawings in DXF, IGES, HPGL and ASCII formats. Attributes are exported in CDF format.

Customization. You can customize fonts, hatch patterns and icon menus. DesignCAD can be programmed with keystroke macros and with the Basic-like programming language.

DesignCAD 3D
American Small Business Computers
One American Way
Pryor OK 74361
(918) 825-9367   Fax (918) 825-6059

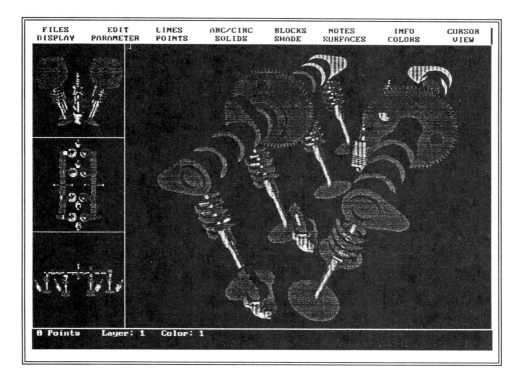

Features. DesignCAD 3D is a 3D-only design package for DOS that performs wireframe and surface modelling. Largest drawing size: unlimited; layers: 62.

File Formats. DesignCAD's proprietary format is called DC3. It imports and exports drawings in DXF. Attributes are exported in CDF format.

Customization. You can customize fonts and hatch patterns. DesignCAD can be programmed with macros.

Drafix Windows CAD
Foresight Resources
10725 Ambassador Drive
Kansas City MO 64153-1298
(800) 231-8574   Fax (816) 891-8018

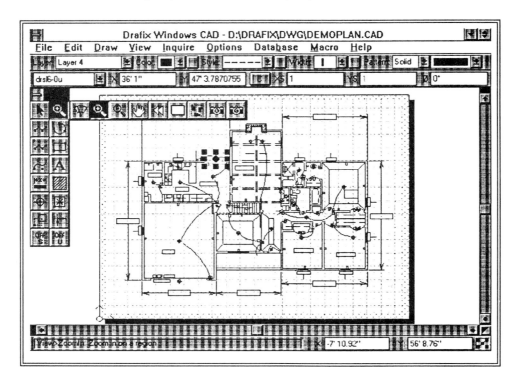

Features. Drafix Windows CAD is a 2D-only drafting package bundled with 400 symbols. Largest drawing size: unlimited; layers: 256.

File Formats. Drafix's proprietary format is called CAD. It imports and exports drawings in DXF, IGES, WMF, and HPGL. Attributes are exported in SDF, CDF, and XLS formats.

Customization. You can customize icon buttons and dialogue boxes with macros and program with the Pascal-like programming language.

EasyCAD
Evolution Computing
437 South 48th Street, Suite #106
Tempe AZ 85281
(800) 874-4028   Fax (602) 967-8633

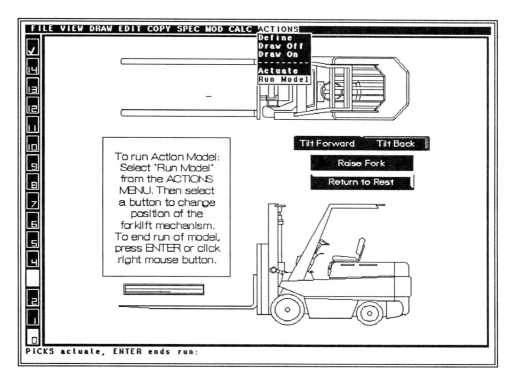

Features. EasyCAD is a 2D-only drafting package for DOS written in assembler for speed. Largest drawing size: unlimited; layers: 256.

File Formats. EasyCAD's proprietary format is called FCD. It imports and exports drawings in DXF format.

Customization. You can customize fonts, linetypes, hatch patterns, and write macros.

FastCAD
Evolution Computing
437 South 48th Street, Suite #106
Tempe AZ 85281
(800) 874-4028   Fax (602) 967-8633

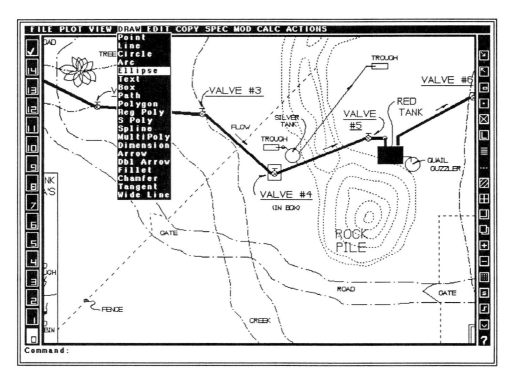

Features. FastCAD is a 2D-only drafting package for DOS written in assembler for speed. Largest drawing size: unlimited; layers: 256.

File Formats. EasyCAD's proprietary format is called FCD. It imports and exports drawings in DXF format.

Customization. You can customize fonts, linetypes, hatch patterns, pull-down menus, write macros, and link programs written in Assembler.

Generic CADD
Autodesk Retail Products
11911 North Creek Parkway South
Bothell WA 98011
(800) 228-3601   Fax (206) 483-6969

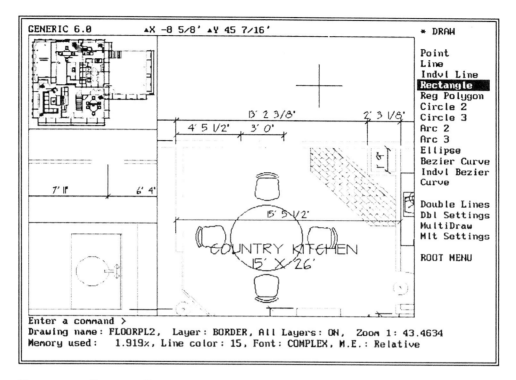

Features. Generic CADD is a 2D-only drafting package for DOS bundled with nearly 500 symbols, AutoCAD-compatible fonts and a bill of material program. Largest drawing size: unlimited; layers: 256.

File Formats. Generic CADD's proprietary format is called GCD. It imports and exports drawings in DWG, DXF, and HPGL formats and several raster formats. Attributes are imported and exported in ASCII, DBF, and WK1 formats.

Customization. You can customize fonts, hatch patterns, screen menus, digitizer buttons, assign macros to all function keys and program with the unnamed programming language.

Generic 3D
Autodesk Retail Products
11911 North Creek Parkway South
Bothell WA 98011
(800) 228-3601   Fax (206) 483-6969

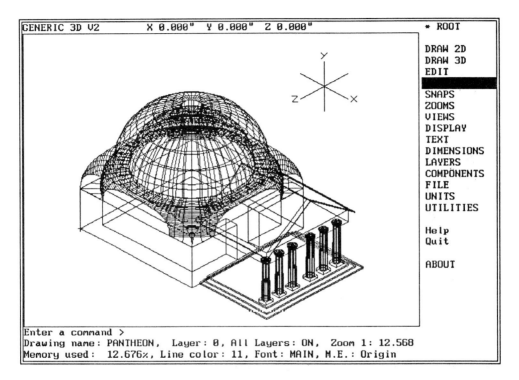

Features. Generic 3D is a 3D-only drafting package for DOS bundled with nearly 500 symbols and AutoCAD-compatible fonts. Largest drawing size: unlimited; layers: 256.

File Formats. Generic CADD's proprietary format is called GCD. It imports and exports drawings in DWG, DXF, and HPGL formats. Attributes are not available.

Customization. You can customize fonts, hatch patterns, screen menus, digitizer buttons, assign macros to all function keys and program with the unnamed programming language.

GFA-CAD for Windows
GFA Software Technologies
27 Congress Street
Salem MA 01970
(508) 744-0201   Fax (508) 744-8041

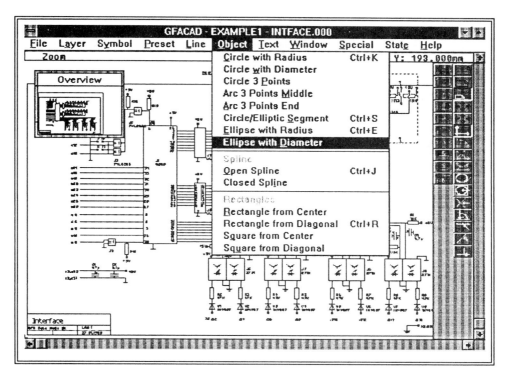

Features. GFA-CAD is a 2D-only layer-oriented drafting package. Largest drawing size: 4MB; layers: 16.

File Formats. GFA-CAD's proprietary format is called INF. It imports and exports drawings in DXF format. Windows Clipboard exports WMF files. Attributes are exported in SDF and DBF formats.

Customization. You can customize hatch patterns, assign HPGL-like macros, supply variable expressions for data input and program with the QuickBasic programming language.

Home Series
Autodesk Retail Products
11911 North Creek Parkway South
Bothell WA 98011
(800) 228-3601   Fax (206) 483-6969

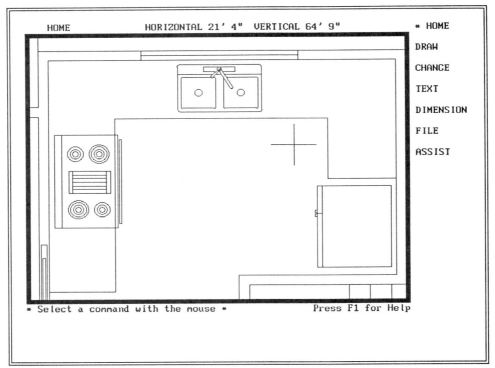

Features. The Home Series is a 2D and 3D drafting package for DOS bundled with symbols. The series is based on a crippled version of an older Generic CADD and consists of separate packages for home, kitchen, bathroom, landscape, deck, and office design.

File Formats. The Home Series' proprietary format is called GCD and is Generic CADD-compatible. It imports and exports drawings in DXF format, exports in DWG format, and imports Generic 3D files.

Customization. You cannot customize any aspect of the Home Series.

Mannequin
HumanCAD
1800 Walt Whitman Road
Melville NY 11747
(516) 752-3568   Fax (516) 752-3507

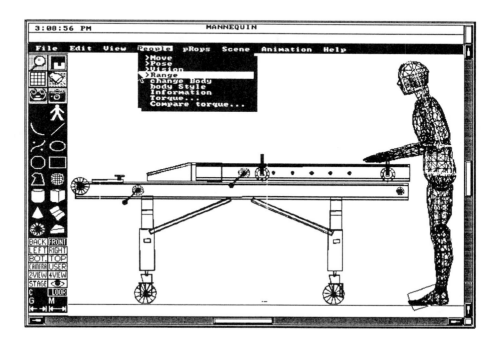

Features. Mannequin is a 3D design package for DOS bundled with nearly 100 symbols. Its primary purpose is to position human figures (of different sizes and nationalities) in any pose. Largest drawing size: unlimited; layers: 1.

File Formats. Mannequin's proprietary format is called HCD. It imports and exports drawings in DXF, and exports drawings 3DS, WMF, and a dozen raster formats.

Customization. You can customize body descriptions and program macros to animate movement.

Mannequin Designer
HumanCAD
1800 Walt Whitman Road
Melville NY 11747
(516) 752-3568   Fax (516) 752-3507

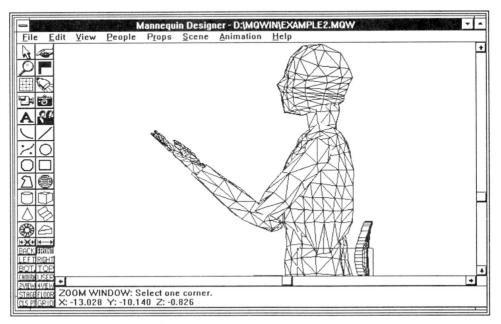

Features. Mannequin Designer is a 3D design package for Windows bundled with nearly 100 symbols. Its primary purpose is to position human figures in any pose. Largest drawing size: unlimited; layers: 1.

File Formats. Mannequin's proprietary format is called MQW. It imports and exports drawings in DXF, and exports in many raster formats.

Customization. You can customize body descriptions and program macros to animate movement.

MicroStation
Intergraph Corp
Huntsville AL 35894-0001
(205) 730-2940   Fax (205) 730-9491

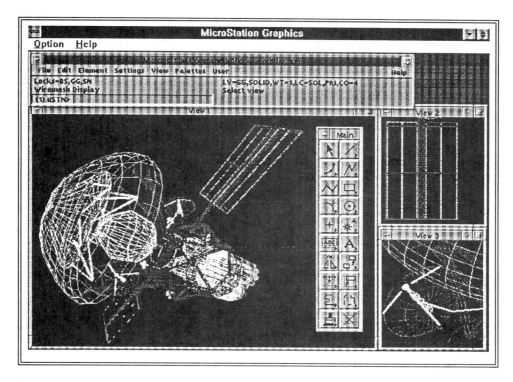

Features. MicroStation is a 2D and 3D drafting package for DOS and Windows bundled with rendering, animation tools, SQL database link and PostScript-like fonts. Largest drawing size: unlimited; layers: 63.

File Formats. MicroStation's proprietary format is called DGN. It imports and exports drawings in DXF, IGES and DWG formats. Attributes are imported and exported in DBF and other database formats.

Customization. You can customize fonts, hatch patterns, screen and tablet menus and digitizer buttons. MicroStation can be programmed in an Assembler-like a Visual C-like programming language. Under Windows, it is an OLE server and has bidirectional DDE.

PC-Draft CAD
Natural Software
19 South Fifth Street
St Charles IL 60174
(708) 377-7320   Fax (708) 506-1808

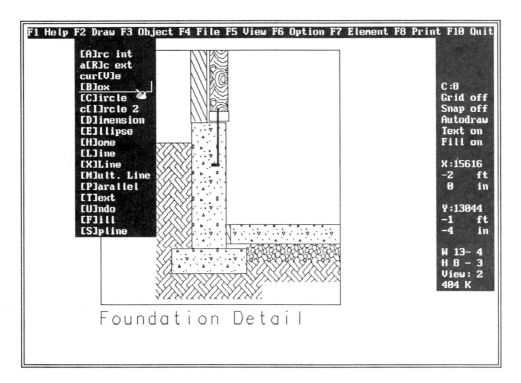

Features. PC-Draft CAD is a 2D-only drafting package for DOS. Layers: unlimited; shareware.

File Formats. PC-Draft CAD's proprietary format is called PCD. It imports and exports drawings in WPG format and imports DXF format files.

Customization. You can customize hatch patterns and write macros. PC-Draft CAD is useful for editing vector-based WPG WordPerfect graphic files.

Planix

Foresight Resources
10725 Ambassador Drive
Kansas City MO 64153-1298
(800) 231-8574   Fax (816) 891-8018

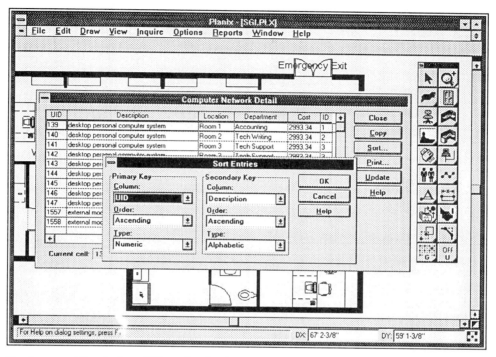

Features. Planix is a 2D-only drafting package bundled with 400 symbols designed for office plans and layouts. Largest drawing size: unlimited; layers: 256.

File Formats. Planix's proprietary format is called CAD. It imports and exports drawings in DWG, DXF, IGES, WMF, and HPGL. Attributes are exported in SDF, CDF, and XLS formats.

Customization. You can customize icon buttons and dialogue boxes with macros and program with the Pascal-like programming language.

3D Concepts for Windows
Autodesk Retail Products
11911 North Creek Parkway South
Bothell WA 98011
(800) 228-3601   Fax (206) 483-6969

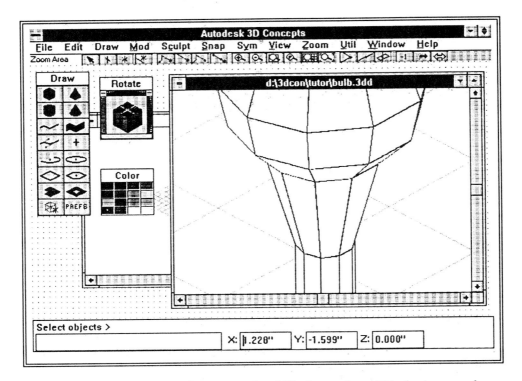

Features. Autodesk 3D Concepts for Windows is a 3D design package bundled with nearly 300 symbols and rendering. 3D Concepts the Windows version of the DOS-based 3D Design companion program to Generic CADD. Largest drawing size: unlimited; layers: 256.

File Formats. 3D Concepts' proprietary format is called 3DD. It imports and exports drawings in 3D DXF and 2D GCD formats. Windows Clipboard exports and imports 3DD format, and exports in BMP and WMF formats.

Customization. You cannot customize 3D Concepts. programming language. OLE and DDE are not supported.

TurboCAD Professional for Windows
International Microcomputer Software, Inc.
1938 Fourth Street
San Rafael CA 94901
(415) 454-7101   Fax (415) 454-8901

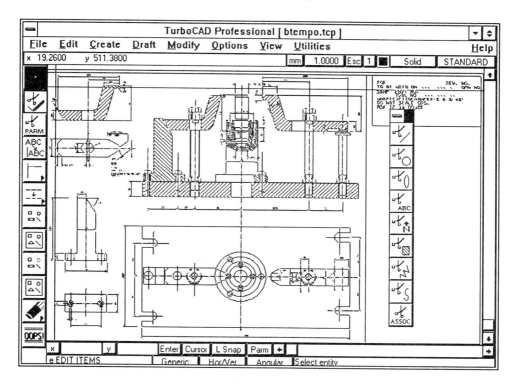

Features. TurboCAD Professional for Windows is a 2D drafting and design package bundled with an on-line bill of material calculator. Largest drawing size: unlimited; layers: 256.

File Formats. TurboCAD's proprietary format is called TCP. It imports and exports drawings in DXF and IGES formats. Attributes are only imported in SDF format. Windows Clipboard is not supported.

Customization. TurboCAD can be programmed with macros, menu definitions, key mappings, and the DPL (drafting programming language) Basic-like programming language. TurboCAD is neither a DDE nor OLE server.

Upfront

Alias Research Inc
110 Richmond Street East
Toronto ON M5C 1P1
(800) 267-8692   Fax (416) 362-4696

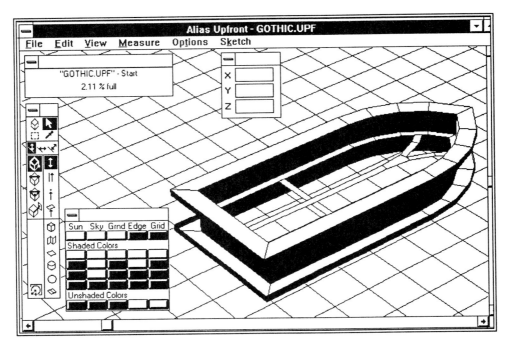

Features. Upfront is a 3D-only design package for Windows bundled with rendering and sun locations for accurate shadow casting. Largest drawing size: unlimited; layers: 252.

File Formats. Upfront's proprietary format is called UPF. It imports and exports drawings in DXF, WMF, ASCII, and several raster formats.

Customization. You cannot customize nor program Upfront.

Vellum 3D for Windows
Ashlar, Inc
1290 Oakmead Parkway
Sunnyvale CA 94086-9669
(408) 746-1800

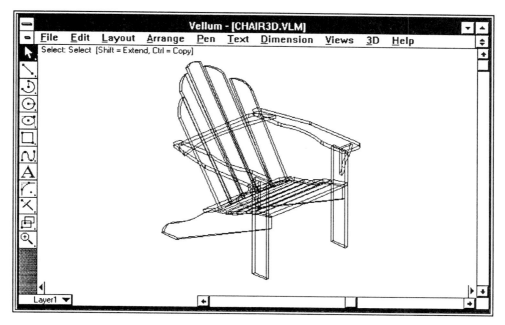

Features. Vellum 3D for Windows is a 3D drafting and design package bundled with 2D analysis, its real-time Drafting Assistant, and parametrics. Largest drawing size: unlimited; layers: 256; uses a hardware lock.

File Formats. Vellums's proprietary format is called VLM. It imports and exports drawings in DXF, IGES, WMF, and BMP formats. Vellum does not have attributes. Windows Clipboard supports WMF, BMP and import of text files.

Customization. You cannot customize Vellum although mathematical functions can be entered in text entry boxes. Vellum does not support DDE or OLE.

VersaCAD Design
Computervision
100 Crosby Drive
Bedford MA 01730
(617) 275-1800   Fax (617) 868-4658

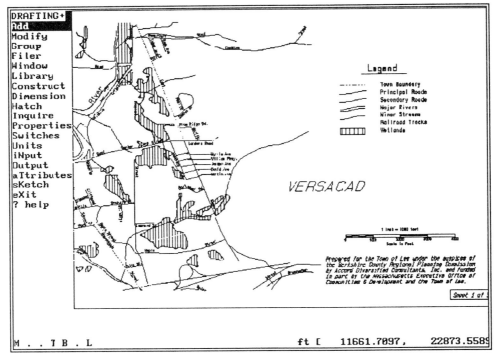

Features. VersaCAD is a 2D and 3D drafting package for DOS bundled with nearly 1,000 symbols, CAD Overlay for raster editing, a plot spooler and drawing translation program. Largest drawing size: unlimited; layers: 256.

File Formats. VersaCAD's proprietary format is called V2D and V3D for 2D and 3D drawings. It imports and exports drawings in DXF and IGES formats. Attributes are exported in ASCII, DBF, and WK1 formats.

Customization. You can customize fonts and linetypes but not hatch patterns. You can assign macros to all function keys and template boxes, and program with the CPL (CAD programming language) Basic- and Pascal-like programming language.

APPENDIX B

CAD Software Sources

There are more than one hundred vendors of CAD packages. The software ranges in price from free (distributed as shareware) to over $70,000 for a system that bundles hardware and software. Most CAD packages a general-purpose; some are specific to analysis or GIS (geographic information system).

Two Hundred and Seventy CAD Packages

The following tables list the names of 270 CAD software packages from 169 vendors, listed alphabetically. CAD packages discussed in this book are shown in *bold italic*; CAD packages listed in the CAD Software Gallery (Appendix A) are shown in *italic*.

The list changes monthly as vendors change the direction of their products or move to a new location. For example, as this book was being finished, Claris announced that they would stop producing ClarisCAD as of 1 October, 1993.

A

Vendor	CAD Product
Abbot, Foster & Hauserman Co (800) 562-0025	3D MiniCAD
Abracadata (503) 342-3030	Design Your Own Home Series Landscape, Architecture, Interiors
Accugraph Corp. (915) 581-1171	MountainTop Advanced Architectural
Acds Graphics Systems (819) 770-9631	AFM MIME
Adra Systems, Inc. (508) 937-3700	CADRA-III
Advanced Relational Technologies Houston TX	Chief Architect Advantage Engineering P&ID Advantage 3D Advantage ElecDes Advantage 2D Advantage
ael-Advanced Graphics Systems, Inc.	VISIONAEL
Aesthesdes NV 011-32-091-218311 Belgium	Aethedes
Alden Computer Systems (508) 744-1314	APPS
Aldus Corporation (800) 685-3514	Aldus IntelliDraw
Algor, Inc. (412) 967-2700	Algor FEA
Alias Research Inc. (416) 362-9181	*Upfront 1.1* Alias Sonata
Alpha Computer Services San Bernardino CA	CivilCADD

Vendor	CAD Product
American Small Business Computers (918) 825-4844	*DesignCAD 2D 6.0* DesignCAD 3D DesignCAD Professional ModelCAD VinylCAD
Amiable Technologies, Inc. (215) 222-9066	FlexiCAD
Anderson Windows Bayport MN	Anderson CAD
APEC Dayton OH	Input Data Manager
Applicon, Inc.	Bravo
Arcadd, Inc (714) 391-4305	Lottacad
Archade O Harris BV 011-31-3480-17590 Netherlands	Arkey-PC
Archway Systems *(714) 374-0440*	*PenDrafter CAD*
Aries Technology (508) 453-5310	ConceptStation
ATS Software 011-65-225-8311 Singapore	Archvisions Supervisions
Artifice, Inc Eugene OR	Design Workshop
ASD International San Francisco CA	P&ID CAD
Ashlar Inc (408) 746-1800	*Vellum 3D* Vellum 2D
Aura CAD/CAM Inc. (213) 536-9207	auraCAD
auto.des.sys, Inc. (614) 488-9777	form.Z

Vendor	CAD Product
Autodesk, Inc. (415) 332-2344	*AutoCAD* AutoSurf/AutoMill
Autodesk Retail Products (800) 228-3601	*AutoSketch for Windows* *Generic CADD 6.1* Generic 3D 3D Concepts for Windows The Home Series Generic CADD Mac
Auto-Graph CDS (606) 269-8585	Auto-Graph
Automated Methods, Inc (415) 750-1976	Ultimate CAD Ultimate CAD 3D Ultimate CAD Mapping ReGIS
Automation Intelligence (407) 661-7000	Expedite 3D/Windows
Auto-Trol Technology Corp. (303) 252-2683	Series 7000 JazzLine

B

Vendor	CAD Package
Baghbagh Montreal PQ	BAGHBAGH CAD
Bearfax Technologies (415) 558-9615	Architron II

C

Vendor	CAD Package
C&G Software Systems Atlanta CA	Land Survey CAD
C.H. Guernsey & Co. (415) 947-5515	Earth One GIS
c.a.s.a.GIFTS, Inc.	GIFTS 6.5.4
Cadam, Inc (800) 255-5710	Micro CADAM Professional CADAM
Cadcorp	wincad
Cadima, Inc.	ASCAE
Cadisys Corp San Jose CA	3D Station
Cadkey, Inc. (203) 647-0220	*CadKey* DataCAD
Cadmax (nee Vector Automation) (410) 532-2803	*CADMAX 5.0*
CadSoft International Guelph ON	AP Design
CADVision International 011-41-71-603-3313 England	X-CAD Professional
CADworks (nee Skok) (800) 866-4CAD CADworks	Drawbase 2000 Drawbase 5000
Camax Systems, Inc. (612) 854-5300	CAMAND ULTRACAM
Computer Integrated Building Corp (707) 874-2826	SolidBuilder
Caroline Informatique	MTEL
Cascade Graphics Systems (714) 757-2972	CADlab Apple IIE CadCAD II/III

Vendor	CAD Package
Cedra Corp. (716) 232-6998	CEDRA Land CEDRA Sand CEDRA Water
Chance & Co 011-619-383-2114 Australia	MacSurf MacHydra
Cimatron 011-972-3-571-5171 Isreal	Cimatron 90
Cimlinc Inc. (800) 225-7943	CIM CAD CIM CAD 3D
Circuit Studios, Inc (705) 522-5335	Velocity
Cisigraph Corp.	STRIM100
CNC Software (203) 875-5006	MasterCAM Draft
Claris Corp (408) 987-7407	Claris CAD
Cognition (nee Automatix) Corp. (508) 667-7900	Mechanical Advantage Design Advantage
COMPUneering, Inc Thornhill ON	FRAME Mac LANDesign LANDview
Computer Design (616) 361-1139	DesignConcept 2D DesignConcept 3D
Computer Easy / Mindware (800) 447-0477	Floorplan Plus for DOS and Windows 3D Design for Windows
Computers and Structures, Inc Berkeley CA	SAP90 ETABS SAFE

Vendor	CAD Package
Computervision (617) 275-1800	*VersaCAD/386* VersaCAD MAC VersaCAD Designer CADDS CVware Parametric Design CV Dimension III DesignView MEDUSA Personal Designer Computervision GIS SYSTEM
Comtran Corp (404) 448-6177	DesignBID
Construction Data Control (800) 221-3929	ProfitCAD
Control Automation Inc (407) 676-3222	Model Mate Plus
Control Data Systems Inc. (612) 482-6736	ICEM DDN ICEM Surf
Creative Visual Software (408) 997-1621	Replimovi Voyager Replimap Repliflow
CSA, Inc Marietta GA	2D Modeller 3D Modeller
CSE Corp (412) 856-9200	Protoll Micro Solids

D

Vendor	CAD Package
Dassault Systems of America (201) 967-2382	Professional CADAM CATIA
Data Automation (619) 693-4070	DGS-2000
Deep River Publishing	Complete House
Deneba Software, Inc.	Canvas 3
Designware, Inc. (617) 924-6715	myHouse
Dickens Data Systems, Inc.	DesignBid
Digitarch USA (800) 633-6564	Flash CAD Magicad
Ditek International (416) 479-1990	DynaCADD for Windows
DP Technology (805) 388-6000	Esprit-X
Dynaware USA (800) 445-3962	DynaPerspective

E

Vendor	CAD Package
EDI Engineering Dynamics Inc Kenner LA	SACS
EDS GDS Solutions	GDS MicroGDS
EDS Unigraphics	Unigraphics
Elcomp Pty (314) 351-2513	Drafteasy

Vendor	CAD Package
Engineered Software (919) 299-4843	PowerDraw
Engineering Systems Corp (800) ESC-CADD	Design Graphix
EnnerCalc Engineering Software Costa Mesa CA	EnnerCalc FastFrame
ESRI, Inc. (714) 793-2853	ARC/INFO
Evans & Sutherland Computer Corp.	CDRS
Evolution Computing (800) 874-4028	*EasyCAD* *FastCAD* FastCAD 3D
Expert Software	Expert Design Series Office, Home, Landscape

F

Vendor	CAD Package
FlexCAD, Ltd Houston TX	IDS-Pipe IDS-Iso IDS-Flow IDS-Steel IDS-Electric
Foresight Resources Corp. (800) 231-8574	**Drafix Windows CAD** *Planix* Drafix CAD Ultra
Forthought, Inc. (803) 878-7484	SNAP!

G

Vendor	CAD Package
Geovision, Inc.	StatMap III Professional
GFA Software Technologies (508) 744-0201	*GFA-CAD for Windows*
GIMEOR, Inc (202) 546-8775	Mac Architron Architron II Architron II PC/PS
Graphisoft (800) 344-3468	ArchiCAD
Graphsoft, Inc. (410) 461-9488	MiniCad+ Blueprint
Ground Modeling Systems, Inc. (206) 670-2520	PANTERRA

H

Vendor	CAD Package
HASP, Inc. (303) 669-6900	Engineering Design System
Hewlett-Packard (303) 229-3303	HP Precision Engineering/ME10 ME 30 HP Precision Engineering/Solid Designer
HumanCAD (516) 752-3507	*Mannequin for DOS* *Mannequin Designer for Windows*

I

Vendor	CAD Package
IBM Corporation (207) 783-7141	Architecture & Engineering Series CAD/One Plus
Icam Technologies (514) 697-8033	AutoCAM
IGC Technology Corp (415) 563-3612	Pegasys I Pegasys II
Imbsen & Assoc Sacramento CA	Highway Bridge Design
International Microcomputer Software (415) 454-7101	*TurboCAD Professional for Windows* TurboCAD for DOS
Infinite Graphics Inc (612) 721-6283	ProCAD
Innovative Data (415) 680-6818	DesignDreams MacDraft
Interactive Computer Modelling (703) 476-1600	ICM GMS ICM Lynx 2 ICM Lynx 3
Intergraph Corp. (800) 345-4856	**MicroStation** IGDS
ISICAD, Inc. (714) 533-8910	**Cadvance for Windows** System 5000

J - L

Vendor	CAD Package
John Keays & Associates	KEAYS Software
Lamutt & Assoc Lakewood CO	Civil Engineering Design
Land Innovation, Inc (612) 420-6811	Site Comp

M

Vendor	CAD Package
MacNeal-Schwendler	MSC/XLplus
Manufacturing and Consulting Services (602) 991-8700	*ANVIL 1000* ANVIL-5000
Masta Corp. Ltd.	Dragon
Matra Datavision Inc.	EUCLID-AEC EUCLID-IS Prelude/Solids
McDonnell Douglas (314) 872-3119	Graphics Design System Unigraphics
MegaCADD (800) 223-3175	MegaMODEL/386 MegaDRAFT
Metrosoft East Rutherford NJ	Robot
Modern Computer-Aided Engineering	INERTIA/InSolid
Murata Wiedemann (704) 875-9280	Control Path

N - O

Vendor	CAD Package
Natural Software (708) 377-7320	PC-Draft-CAD
OrCAD (503) 690-9881	OrCAD

P

Vendor	CAD Package
PacSoft Kirkland WA	Cogo Road and Site Design
Palette Systems Inc (603) 866-1230	Palette
Parametric Technology Corp. (617) 894-7111	Pro/ENGINEER
Pathtrace Systems (714) 595-6767	PAMS
PDA Engineering	PATRAN 3
Pella Windows & Doors Pella IA	Designer CAD
Pelton Engineering (604) 758-7726	ECAD
Plus III Software Inc Atlanta GA	TerraModel
Point Line (nee Robi Graphics) (608) 831-0077	Point Line Professional CADD

R

Vendor	CAD Package
RDS Systems, Inc (301) 984-1989	MAPS
Research Engineers Orange CA	STAAD-III AutoCivil
RISA Technologies Lake Forrest CA	RISA-3D QuickConnect
Robo Systems International (800) 221-ROBO	RoboDesign RoboCAD RoboSolid

S

Vendor	CAD Package
Sayet Corp (801) 944-9212	LaserCAD Professional LaserCAD Apprentice
Shamrock Systems (504) 752-7250	Advanced Shamrock CAD Professional Shamrock CAD
Schroff Development Corporation (913) 262-2664	SilverScreen III
Sigma Design, Inc. (617) 890-4904	ARRIS
Silicon Beach Software (619) 695-6956	Super 3D
Simcus (nee Trilex) 011-44-81-546-9670 England	WP Illustrator
Software Associates (415) 367-4994	PowerCAM PowerCAM Lite
Solutionware (408) 249-1529	Geopath CAM/CAM
Stephen Dedalus (nee Euclidean Systems) (919) 467-7072	MetaSite MetaPlan
StereoCAD, Inc.	REALTIME
Strata, Inc.	StrataVision 3D
Strategic Mapping, Inc.	Atlas GIS
Structural Dynamics Research Corp. Milford OH	I-DEAS
Swanson Analysis Systems, Inc.	ANSYS FEA
Synthesis Company Bellingham WA	Synthesis

T

Vendor	CAD Package
Tangram Computer Aided Engineering	SWIFT
Teksoft (602) 942-4982	ProCAD 3D/Surf/Mill ProCAD/CAM
Tripod Data Systems Corvallis OR	Easy Survey
Trius, Inc (508) 794-9377	DraftChoice

U - Z

Vendor	CAD Package
UNIC, Inc.	Architrion II
Varimetrix Corp. (407) 676-3222	Varimetrix Drafting Varimetrix Modelling Model Mate Plus
Virtus Corp (919) 467-9700	Virtus Walkthrough
Wavefront Technologies, Inc. (805) 962-8117	The Advanced Visualizer
West Coast Consultants (619) 565-1266	Curve Digitizer CAD
Wiechers & Partners 011-49-2173-39640 Germany	LogoCAD

APPENDIX C

CAD Layer Standards

The problem of defining a standard for naming CAD layers is difficult enough that multiple standards have evolved. This appendix presents two standards in somewhat greater detail than Chapter 4, "Colors and Layer Names": one created by the AIA and the other from FDM and based on the CSI.

American Institute of Architects

The AIA layer standard uses up to five fields to synthesize a layer name. For example, the layer `ETLGHT-EMER-XXXX` contains the temporary emergency lighting. The layer name is created, as follows:

AIA Layer Name Format

Field	Group	Description
1	*Example:* E	Construction Category
2	*Example:* ET	Modifier
3	*Example:* LGHT	Major Group
4	*Example:* EMER	Minor Group
5	*Example:* XXXX	User Defined

Construction Category

The AIA defines eight construction categories:

Construction Categories

Category	Description
A	Architectural, interior, facilities
E	Electrical
F	Fire protection
L	Landscape, site and civil
M	Mechanical
P	Plumbing
S	Structural
X	Information

Modifier

The standard defines five modifiers:

Layer Modifier

Modifier	Description
E	Existing to remain
F	Future construction
S	Shell or base building
T	Temporary construction
X	Existing to be demolished

Major Group

There are a large number of major groups. For example, the A (architectural) construction category uses the following major groups:

A-Category Major Groups

Group	Description
CLNG	Ceiling
DETL	Details
DOOR	Doors
ELEV	Elevations and 3D surfaces
EQPM	Equipment
FLOR	Floors
FURN	Furniture
GLAS	Windows and glazed partitions
ROOF	Roof edge and features
SCTN	Sections
WALL	Walls, curtain walls, storefront

Minor Groups

Each major group has one or more minor groups. For example, the CLNG (ceiling) major group has the following minor groups:

CLNG-Group Minor Groups

Group	Description
BKHD	Bulkhead
GRID	Grid
OPEN	Ceiling-roof penetration
TEES	Main tees

User Defined
The four-character user-defined field can be used for any purpose.

Three Naming Choices
The AIA created a three naming systems to accommodate the layer naming capabilities of different CAD packages. The three systems are suitable for AutoCAD's 31-character layer names (called the "long-form" name), CAD packages with eight-character layer names (called the "short-form" name), and those limited to numbered layers, called the "numbered" name.

- The long-form name has four or five fields: E-LGHT-EMER-XXXX or ETLGHT-EMER-XXXX

- The short-form name consists of the first two fields: E-LGHT or EXLGHT

- The numbered name is assigned to specific layer names: 130 for the emergency lighting layer

For More Information
For a complete copy of the AIA layer guidelines, send $20 to The American Institute of Architects, 1735 New York Avenue NW, Washington DC 20006.

Facilities Data Management
The 16DIV layer protocol is based on the CSI (Construction Specification Institute) MasterSpec divisions. The layer standard uses three fields to synthesize a numerical layer name. For example, the layer 16531 contains the existing emergency lighting to remain. The layer name is created, as follows:

FDM Layer Name Format

Field	Group	Description
1	*Example:* 16	MasterSpec Division
2	*Example:* 53	Division
3	*Example:* 1	Modifier

MasterSpec Division

The CSI defines 16 construction categories:

The 16 MasterSpec Divisions

Division	Description
1	Specific
2	Civil, Landscape
3	Concrete
4	Masonry
5	Metals
6	Wood, plastics
7	Thermal, moisture protection
8	Doors, windows
9	Finishes
10	Specialties
11	Equipment
12	Furnishing, CAD specific
13	Special construction
14	Conveyor systems
15	Mechanical
16	Electrical

Division

The standard defines many divisions within each MasterSpec division. Here are the Division 16, Electrical, divisions:

Division 16, Electrical

Divisions	Description
11	Raceway
12	Wiring, circuiting
13	Device
16	Cabinet
21	Generator
31	Substation
32	Transformer
42	Switch gear
43	Meter
51	Lighting fixture
53	Emergency fixture
63	Battery power system
72	Alarm, detection system
73	Data cable system
74	Telephone system
75	Nurse call system
76	Intercom system
77	Public address system
78	Television system
85	Electrical resistor heating

Modifier

The last character in the layer name modifies the objects in the division:

Layer Modifier

Modifier	Description
1	Existing to remain
2	Existing to be removed
3	New
4	Text
5	Dimension
6	Hatch pattern
7	*User definable*
8	*User definable*
9	*User definable*

APPENDIX D

The AutoCAD Certification Exam

Professional certification has long been an important way for professionals, including accountants, attorneys, doctors, and architects, to demonstrate that they have the skills and knowledge necessary to do the job. It provides a consistent and fair standard against which you and your employees can evaluate abilities.

With the AutoCAD Certification exam, you can determine expertise with AutoCAD software. Drafters gain a competitive advantage when they apply for a job.

What is the AutoCAD Certification Exam?

The currently available certification exam tests for AutoCAD fundamentals. The exam measures standard knowledge and skills of the professional who has completed at least 48 hours of intensive AutoCAD instructions or who has used AutoCAD regularly for several months. It is also appropriate for new graduates of credit CAD drafting programs at community colleges, technical schools, or other training centers.

The Format of the Exam

The AutoCAD Certification exam is divided into two levels: Level I (Beginner) and Level II (Advanced). Each level is divided into two parts: a general knowledge test and a drawing test. The first part of the exam is a two-hour, hands-on set of five drawing exercises using AutoCAD software on a computer. You will be required to create drawings of varying difficulty. Multiple-choice questions accompany each drawing and test your understanding of AutoCAD features. For example, you may be expected to determine locations and relationships among drawing entities using the inquiry commands.

The second part of the exam is a one-hour multiple-choice section. The 75 questions are comprehensive and test for a thorough knowledge of AutoCAD commands and features. Below is a chart of the question categories and their weighting for the Level I examination.

Level I AutoCAD Certification Exam Format

Multiple Choice Categories	Percentage of Score	Number of Questions
General AutoCAD Terminology	5%	4
User Coordinate Systems	4%	3
Inquiry Commands	5%	4
Beginning Plotting	4%	3
Beginning Selection Sets	5%	4
Draw Commands	19%	14
Edit Commands	23%	16
Beginning Grips	3%	2
Beginning Dimensioning	5%	4
Block Commands	5%	4
Settings Commands	8%	6
Layers	5%	4
Display Commands	5%	4
Beginning Utility Commands	4%	3
Total	100%	75

The examination is electronic: a computer loaded with the current release of AutoCAD is provided. Both parts of the test are closed book. Test takers are not permitted to use any printed materials, scratch pads, or paper during the exam; calculators are also not permitted.

How to Prepare for the Exam

To prepare for the exam, review AutoCAD menus and commands. Having a basic understanding of DOS or other operating systems is strongly recommended. Consider attending a course at an Autodesk Training Center. Or you can purchase the *AutoCAD Certification Examination Preparation Manual* (by Alan Kalameja, Delmar Publishers; $32.95), a self-study workbook to help you with exam preparation for both Levels.

For More Information

To register for the AutoCAD Certification exam, contact:

Drake Training and Technologies
8800 Queen Avenue South
Bloomington MN 55431
(800) 995-EXAM

Price for each Level of the examination is $150.00. For more information about Autodesk Training Centers, contact:

Autodesk Education Department
2320 Marinship Way
Sausalito CA 94965
(800) 964-6432

The following pages contain sample exam questions. Answers are at the end of this appendix. *Reprinted with permission of Autodesk, Inc.*

Drawing Exercises

Problem 1.
Directions: Create the drawing of the object shown in Figure 1 according to the following instructions.

1. Using the Units command:
   a. Set units to decimal.
   b. Set the number of digits to the right of the decimal point to two.
   c. Accept the defaults for the remaining prompts.

2. Omit the dimensions.

3. Begin your drawing with the 1.25 diameter circle using the coordinates (5,5) as the center point.

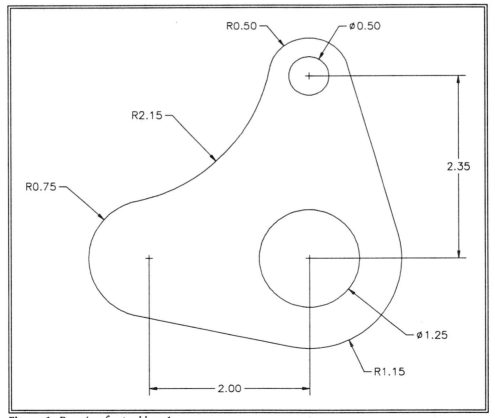

Figure 1: *Drawing for problem 1.*

4. The coordinates of the 0.50 diameter circle are (5.45, 7.35).

5. When you have finished the drawing, answer each of the questions 1 and 2 by selecting the best choice of answers given.

Questions.
1. The x,y-coordinates of the center of the 2.15-radius arc are:
   A. 2.43,6.81
   B. 2.73,7.26
   C. 2.81,7.84
   D. 2.86,7.90

2. The net area of the object with the two holes removed is closest to:
   A. 7.75
   B. 7.77
   C. 7.79
   D. 7.81

Answers are at the end of this appendix.

Problem 2.

Directions: Create the drawing of the outline of the house shown in Figure 2 according to the following instructions.

1. Using the Units command:
   a. Set units to architectural.
   b. Accept the defaults for the remaining prompts.

2. Set the text style and font to standard Complex.

3. Set the following dimension variables to the values shown.

Dimvar	Value
DimAsz	2'
DimDli	6'
DimExe	2'
DimExo	1'
DimTxt	2'

4. You need draw only the outline of the house and the dimensions.

5. Begin your drawing with the lower left corner of the house (indicated by A) located at the coordinates (10'-0",10'-0").

6. When inserting the vertical dimensions, use the Continue option and locate them at an x-coordinate value of 105'-0".

7. When inserting the horizontal dimensions, they are to be inserted using the Baseline option with the 26' dimension to be located at a y-coordinate value of 70'-0".

8. When you have finished the drawing, answer questions 3 and 4 by selecting the best choice of answers given.

Questions.
3. The distance from the corner at B to the intersection of the arrowhead and extension line at D is closest to:
    A. 85'-2"
    B. 85'-5"
    C. 85'-8"
    D. 85'-11"
    E. 85'-2"

4. The distance from the intersection of the arrowhead and the extension line at C to the corner of the house at A is closest to:
    A. 115'-1"
    B. 115'-5"
    C. 115'-9"
    D. 116'-1"
    E. 116'-5"

Answers are at the end of this appendix.

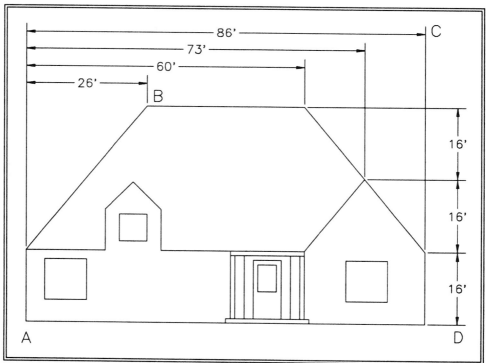

Figure 2: *Drawing for problem 2.*

Problem 3.

Directions: Create the drawing of the object shown in Figure 3 according to the following instructions.

1. Using the Units command:
   a. Set units to decimal.
   b. Set the number of digits to the right of the decimal point to four.
   c. Accept the defaults for the remaining prompts.

2. Omit the dimensions.

3. The three sides of each tab are composed of an arc and parallel lines 10.00 apart.

4. The two holes of diameter 10 are located at a radius of 60.

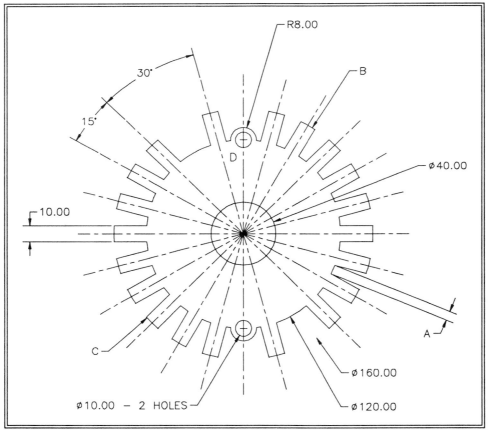

Figure 3: *Drawing for problem 3.*

5. Begin your drawing with the 40-diameter circle using the coordinates (140,120) as the center point.

6. When you have finished the drawing, answer each of the questions 5, 6, 7, and 8 by selecting the best choice of answers given.

Questions.

5. Move the entire drawing a polar distance of 12 at an angle of 6 degrees. The new x,y-coordinates at the center of the 10-diameter circle at D are:
   A. 151,9343, 181.2543
   B. 149.9452, 181.0453
   C. 154.9178, 181.5679
   D. 151.8832, 181.6701
   E. None of the above

6. What is the distance *along the arc* indicated by A?
   A. 5.6506
   B. 5.6713
   C. 5.6964
   D. 5.7136
   E. None of the above

7. The distance from the corner B to the corner C is:
   A. 158.8342
   B. 158.6312
   C. 159.6263
   D. 159.6872
   E. None of the above

8. What is the surface area of the plate with the three holes removed?
   A. 14,106.2703
   B. 14,106.2508
   C. 14,106.2316
   D. 14,106.2157
   E. None of the above

Answers are at the end of this appendix.

Multiple-Choice Questions

Directions.
For each of the questions in this section, choose the best choice of answers given.

1. The Setvar command is used to change:
    A. Global variables
    B. System variables
    C. Real variables
    D. Integer variables
    E. Local variables

2. The Continue subcommand of Dim always:
    A. Offsets each added dimension
    B. References the variable DimDli
    C. Prompts for the first extension origin
    D. Is used as part of an angular dimension
    E. Requires an existing linear dimension

3. The Dxfin command converts:
    A. An ASCII file
    B. A BAK file
    C. An IGES file
    D. A DWG file
    E. An EXE file

4. AutoCAD's drawing database retains at least:
    A. 14 significant figures
    B. 16 significant figures
    C. 18 significant figures
    D. 20 significant figures
    E. 22 significant figures

5. The Insert command can be used to:
    A. Delete a block
    B. Create a block
    C. Save a block
    D. Insert any drawing file
    E. Redefine a block's insertion point

6. A backup file is updated when using the commands:
   A. Quit or Save
   B. Files or End
   C. End or Save
   D. Shell or Files
   E. Quit or End

7. When plotting, pen numbers are assigned to:
   A. Colors
   B. Layers
   C. Elevations
   D. Thicknesses
   E. Linetypes

8. The Pan command cannot be used transparently:
   A. If a Line command is in progress
   B. When a regeneration is required
   C. When the fast zoom mode is on
   D. While in the dimension mode
   E. While the snap mode is on

9. The Change command lets you modify:
   A. A dimension
   B. A polyline width
   C. A point figure size
   D. Text
   E. A filename

10. A color cannot be assigned to:
    A. A layer
    B. An entity
    C. A polyline
    D. A block
    E. a Linetype

Answers are at the end of this appendix.

Answers

Drawing Exercises
1. D
2. B

3. A
4. D

5. A
6. C
7. B
8. B

Multiple-Choice
1. B
2. E
3. A
4. A
5. D

6. C
7. A
8. B
9. D
10. E

Index

Index

A

Adding attributes to symbols 52
Add-ons determining primary application 15
Additional charges, service bureau 171
Anatomy of a dimension 73
Anvil-1000MD software 212
Appointing symbol layers 51
Archiving drawings 156
Arrowheads 76
AutoCAD for DOS software 92, 213
   certification exam 263
  for Windows software 214
Autosketch for DOS software 215
  for Windows software 136, 216

B

Benefits of a network 192
Better drawing accuracy 3
Bureau services 169

C

CAD (computer-aided design)
    database accuracy 181
    layer standards 255
    manager, role of 1
    software sources 239
    systems, limitations of 45
    writing standards 143
Cadkey software 101, 217
Cadmax Truesurf software 218
Cadvance for Windows software 107, 219
Caltrans
    filename convention 57
    layer conventions 44
Central file server for networks 189
Changing the organization 7
Claims by vendors 13
Client presentations 5
Color
    and layer names 33
    how CAD works with 36
    the case for color 35
    the case for monochrome 33
    why use in drawings? 33
Communication 8
Competitive pressure 24
Components of a network 188
Conflicting disciplines 46
Contract considerations, service bureau 171
Controlling management expectations 21
Converting drawings 157
Cost
    estimates 4
    of CAD 1
    of revision 28
    money 5
Creating layers 83

D

Data exchange   173
Database links   203
Deciding symbol orientation   50
Description-floor-discipline-sheet filename   56
Design analysis   4
DesignCAD software
   3D software   221
   2D software   128, 220
Determining symbol scale   50
Digitizing drawings rates, service bureau   170
Dimensions   71
   anatomy   73
   arrowheads   76
   dimension line   75
   element colors   78
   extension lines   75
   parameters   74
   prototype drawing   84
   scaling   78
   text   77
Discipline-type-detail-sheet-revision filename   56
Documenting
   standards   143
   symbols   53
Drafix Windows CAD software   114, 222
Drawing symbols   49
Drawings,
   Caltrans filename convention   57
   exchange file formats   175
   exchanging   173
   exercises   266
   file names   55
   file extensions   59
   horror of translation   179
   named by
      description-floor-discipline-sheet   56
      discipline-type-detail-sheet-revision   56
      externally referenced drawings   56
      externally referenced drawings   56

Drawings, *continued*
  named by, *continued*
    project-discipline-drawing 55
  translating 179
  origin and extents 83
  preparing the prototype 81
  uses for external drawings 54
Dual-CAD office 14
DWG (AutoCAD and CadMax drawing file) 178
DXF (drawing interchange format) 175
  via AutoCAD 183
  via third-party software 185

E

EasyCAD software 223
Education 6
Electronic drawing 153
Eliminating paper trail 155
Entity limits 181
Examination, AutoCAD certification 263
Extension lines 75
Externally referenced drawings 56

F

FastCAD software 224
Faster drawing production 3
Fearing the unknown 3
File names 47, 55
  Caltrans filename convention 57
  description-floor-discipline-sheet 56
  discipline-type-detail-sheet-revision 56
  drawing exchange 175
  externally referenced drawings 56
  file extensions 59
Fill turned off 65

Final analysis   29
Fonts   61
   fill turned off   65
   layers turned off   65
   linetypes and hatch patterns   61
   prototype drawing   83
   quick text   65
   scaling   69
   substitution   65
   translating text entities   181
Forgetting the training   26

G

Gallery of CAD software   211
Generic CADD software   121, 225
   3D software   226
GFA-CAD for Windows software   227
Giving engineers the tools   27

H

Hardware   11
   network   187
   selecting   11, 16
   upgrading   20
Hardwired vs custom linetypes   66
Hatch patterns   61, 68
   prototype drawings   84
   scaling   69
Home Series software   228
Horror of drawing translation   179
How
   CAD works with colors   36
   to create a symbol library   49
   to name a layer   38

HPGL (Hewlett-Packard graphics language) 177
Human quirks 46

I

IGES (initial graphics exchange format) 176
Input 153
   input options 156
   service bureau 169

L

LAN (local area network) 187
Layers 33
   conflicting disciplines 46
   how to name 38
   human quirks 46
   in CAD 38
   limitations of CAD systems 45
   name standards 45, 255
   names 33
   naming 38
   off 65
   prototype drawings 83
   standards 45, 255
   strategies
      #0: do nothing 39
      #1: the simple plan 39
      #2: the plotter plan 39
      #3: the four-step plan 40
      #4: do what your client says 42
   symbol layers 51
   what are layers? 38
Limitations
   of CAD systems 45
   resources 8

Linetypes 61, 66
    hardwired vs customized 66
    one-dimensional vs two-dimensional 67
    prototype drawing 83
    scaling 67, 69
    software vs hardware 67
Load
    external references 84
    hatch patterns 84
    linetypes 83
    symbol libraries
    text fonts 83

M

Macros and scripts 200
Maintenance, service bureau 169
Management
    controlling expectations 21
    keeping in loop 28
    LAN manager 191
    not spending on training 25
    role of CAD manager 1
    sending managers to trade shows 24
    training 26
Mannequin software 229
    Designer software 230
Market competition 5
Maximizing CAD efficiency 195
Microstation software 85, 231
Money 5
Monthly user meetings 27
Multiple LANs 190

N

Names
  drawings 55
  extensions 59
  files 55
  layers 38
  prototype drawings 84
  symbols 52
Networking CAD 187
  cable 189
  interface card 189
  operating system 188
  post tips on 25
  traffic cop functions 191
  styles 189

O

Office standards manual 143
On-line help 26
Open a new drawing 83
Operators
  examination 263
  keeping informed 9
  support staff 7
Optimizing CAD 206
Organization, changes 7
Output 153
  options 162
  service bureau 169
Overcoming problems with CAD 21

P

Parametric symbol libraries   197
Patterns, hatch   68
   hatch scaling   69
   prototype drawings   84
PC-Draft CAD software   232
Peer-to-peer network   190
Planix software   233
Planning transition to CAD   9
Plotting   162
   prototype drawing   84
   service bureau rates   170
   solving problems   163
Preparing
   for CAD   9
   prototype drawing   81
Problems
   created by CAD   5
   overcoming   21
Programming languages   201
Project-discipline-drawing filename   55
Proprietary drawing files   173
Prototype drawings   81
   angular measurement   83
   dimension variables and scale   84
   drawing modes   83
   external references   84
   hatch patterns and scale   84
   layers   83
   linetypes and scale   83
   make backups   84
   object snaps   83
   open 83
   origin and extents 83
   patterns and scale   84
   plotter and set plot style   84
   preparing   81
   resolution of measurement   83
   save   84
   symbol libraries   84

Prototype drawings, *continued*
    text fonts and define text styles 83
    unit of measurement 83

Q

Quick text 65

R

Reasons to adopt CAD 3
Return on investment 23
Rules
    "Anything drawn twice should be turned in a symbol" 50
    "Avoid proprietary solutions; buy standard parts" 15
    "Don't buy anything until its need is proven" 20
    "Keep everyone fully informed" 5
    "Only upgrade when the need is proven" 28
    "Paperless office is as likely as a paperless bathroom" 155
    "Purchase standards, not state-of-the-art" 19
    "Replace the CPU when the speed improvement is 3x" 20
    "Software determines the hardware" 15
    "Use the CAD software your clients use" 12

S

Save prototype drawing 84
Scaling 69
    dimensions 78
    fonts 69
    hatch patterns 69
    linetypes 69
    linetypes 67

Screen and tablet menus   199
Selecting
   CAD hardware   16
   symbol insertion point   50
   default plotter   84
   software and hardware   11
Sending managers to trade shows   24
Service bureau pricing   170
Set
   dimensions   71
   dimension variables   84
   object snaps and drawing modes   83
   resolution of measurement   83
   style of angular measurement   83
   unit of measurement   83
Software
   CAD vendors sources   239
   gallery of CAD   211
   selecting   11
   vs hardware linetypes   67
Sources of symbols   54
Speed
   CAD system   195
   optimizing   206
   text   64
Standards
   colors   33
   dimensions   71
   file names   55
   fonts   61
   hatching   68
   layers   38, 255
   linetypes   66
   office library   148
   patterns   68
   prototype drawing   81
   symbols   47
   text   61
   writing   143
Store the symbol   53

Strategies for drawing names
   #0: do nothing   39
   #1: the simple plan   39
   #2: the plotter plan   39
   #3: the four-step plan   40
   #4: do what your client says   42
Styles
   dimension   79
   font   61
   linetype   66
   text   61
Support staff   7
Symbols   47
   adding attribute information   52
   appointing the layers   51
   deciding the orientation   50
   determining the scale   50
   documenting   53
   drawing the symbol   49
   external references   54
   filenames   47
   how to create   49
   in CAD   50
   naming   52
   parametric libraries   197
   prototype drawing   84
   selecting insertion point   50
   sources   54
   storing   53
   what are symbols?   47

T

Text   61
   dimension   77
   fill turned off   65
   font substitution   65
   fonts and styles   61
   layers turned off   65

Text, *continued*
   prototype drawing   83
   quick text   65
   speed   64
   translating text entities   181
3D Concepts for Windows software   234
Training   26
   forgetting about   26
   management   26
   service bureau rates   170
Transition to CAD   9
Translation   173
   database accuracy   181
   DXF via AutoCAD   183
   entity limits   181
   test grid   185
   text entities   181
   unique entities   179
TurboCAD Professional for Windows software   235

U

Unique entities   179
Upfront software   236
Upgrading hardware   20
Uses for external drawings   54

V

Vellum 3D for Windows software   237
VersaCAD Design software   238

What is a?
  layer   38
  network   187
  prototype drawing   81
  service bureau   167
  symbol   47
When management doesn't spend   25
Why?
  CAD   1
  customize CAD   195
Windows   204
WMF (Windows metafile)   177
Working with a service bureau   167
Writing the CAD standard   143

-NOTES-

-NOTES-

-NOTES-

-NOTES-

-NOTES-

-NOTES-